AF856758

Raspberry Pi 5 for Radio Amateurs

Program and Build Raspberry Pi 5 Based Ham Station Utilities with the RTL-SDR

Dogan Ibrahim, G7SCU
Ahmet Ibrahim, 2E1GUC

elektor
design > share > earn

This is an Elektor Publication. Elektor is the media brand of
Elektor International Media B.V.
PO Box 11, NL-6114-ZG Susteren, The Netherlands
Phone: +31 46 4389444

Declaration

The authors and publisher have used their best efforts in ensuring the correctness of the information contained in this book. They do not assume, or hereby disclaim, any liability to any party for any loss or damage caused by errors or omissions in this book, whether such errors or omissions result from negligence, accident or any other cause.

British Library Cataloguing in Publication Data
A catalogue record for this book is available from the British Library

ISBN 978-3-89576-612-1 Print
ISBN 978-3-89576-613-8 eBook

www.elektor.com
Editor: Jan Buiting, MA
Prepress Production: D-Vision, Julian van den Berg
Printers: Ipskamp, Enschede, The Netherlands

Contents

Preface

In recent years there have been major changes in the radio equipment used by the radio amateurs. Although the classical HF and mobile equipment is still in use by large numbers of amateurs, computers and digital techniques are becoming very popular among amateur radio operators or 'hams'. In early days of digital communications, personal computers were used by hams to communicate with each other. PCs have the disadvantage that they are rather expensive and bulky. Nowadays, anyone can purchase a £40 Raspberry Pi computer and run almost all of the amateur radio software on this computer, which is slightly bigger than the size of a credit card.

The Raspberry Pi 5 is the latest credit-card sized computer from the Raspberry Pi Foundation that can be used in many applications, such as in audio and video media centers, as a desktop computer, in industrial controllers, robotics, and in many domestic and commercial applications. In addition to many features found in other Raspberry Pi computers, the Raspberry Pi 5 offers Wi-Fi, Classic Bluetooth, and Bluetooth BLE capability which makes it highly desirable in remote and Internet-based control and monitoring applications.

The Raspberry Pi 5 is a 64-bit quad-core Arm Cortex-A76 processor running at 2.4 GHz, which is two to three times the performance boost when compared to the earlier Raspberry Pi 4. The Raspberry Pi 5 comes with an enhanced graphic performance, using the 800-MHz VideoCore VII graphics chip. Additionally, the Raspberry Pi 5 features a Southbridge chipset made for the first time by the Raspberry Pi Foundation. With the help of this RP1 Southbridge, the Raspberry Pi 5 delivers higher performance and functionality for peripheral devices. It should now be possible to carry out many real-time operations such as audio digital signal processing, real-time digital control and monitoring, and many other real-time operations using this tiny powerhouse.

The RTL-SDR devices (V3 and V4 dongles) have become very popular among radio fans because of their very low cost (some around £12) and rich features. A basic system may consist of a USB-based RTL-SDR device (dongle) with a suitable antenna, a Raspberry Pi 5 computer, an USB-based external audio input/output adapter, and software installed on the Raspberry Pi 5 computer. With such a simple setup, it is possible to receive signals from around 24 MHz to over 1.7 GHz.

This book has four purposes. Firstly, it is aimed to teach the installation of the operating system and basic operating principles and features of the Raspberry Pi 5 to beginners. Secondly, many hardware-based projects are given using the Raspberry Pi 5 together with the Python programming language. These projects have been chosen to be useful to amateur radio operators. Thirdly, the book explains in some detail how to use the RTL-SDR devices (both V3 and V4) together with a Raspberry Pi 5 and popular RTL-SDR software to tune in and receive signals from a wide range of ham and other frequency bands. Lastly, the book also explains how to install and use some of the popular amateur radio software packages on the Raspberry Pi 5.

It is important to realize that the book uses the Python programming language on the latest Raspberry Pi 5 platform, and most of the programs in the book will not work on older versions of Raspberry Pi computers. Readers interested in exploring older Raspberry Pi models in amateur radio projects are recommended to purchase author's earlier book entitled: *Raspberry Pi for Radio Amateurs: Program and build RPI-based ham station utilities, tools, and instruments* (available from Elektor).

I hope you enjoy reading the book.

Dogan Ibrahim, G7SCU and Ahmet Ibrahim, 2E1GUC
London, 2024

Chapter 1 • Installing the Raspberry Pi 5 Operating System

1.1 Overview

The Raspberry Pi 5 is the latest credit card size computer from the Raspberry Pi Foundation. It is based on 2.4-GHz Cortex-A76 Arm processor with a new Southgate bridge for handling the peripheral interface. A new VideoCore VII GPU is provided with 800 MHz speed. A dual camera interface is another nice feature of the Raspberry Pi 5. The microSD card interface supports cards that work at much higher speeds than previously.

The Raspberry Pi 5 is similar to the older Raspberry Pi 4, where both devices have dual 2 4Kp60 HDMI display interfaces, although the Pi 5 supports HDR output. The 2×20-pin GPIO interface is the same in both devices. The Raspberry Pi 5 additionally supports two camera interfaces, a PCIe bus connector, a UART interface, an RTC clock power connector, and a fan power connector. Wi-Fi and Bluetooth are supported by both devices. The on-board power switch on the Raspberry Pi 5 is a useful feature. The Raspberry Pi 5 is powered from a 5 V, 4 A USB-C type power supply and is slightly more expensive than the Raspberry Pi 4.

The camera and display connectors on Raspberry Pi 5 are 15-pin and smaller instead of the original 22-pin connector used on Pi 4. A ribbon cable with a 22-pin connector on one side and a 15-pin one on the other side is required to connect an existing Raspberry Pi 4 camera to Raspberry Pi 5. The Raspberry Pi 5 has two connectors, allowing two cameras or DSI displays (or a mix of) to be connected. The PCIe connector is for fast external PCIe compatible peripherals, such as SSDs.

The new power button on Raspberry Pi 5 could be very useful. When the device is ON, pressing the button brings up the shutdown (logout) menu. A safe shutdown will occur with another press of the power button.

Figure 1.1 shows the front view of the Raspberry Pi 5 with the components labelled for reference.

Figure 1.1: The Raspberry Pi 5.

The Raspberry Pi 5 can get very hot when working. While the 'idle' CPU temperature is around 50 degrees Celsius, it can go higher than 85 degrees under stress. It is recommended to use a cooler to lower the CPU temperature. A dedicated active cooler is available for Raspberry Pi 5. Holes and power points are provided on the board to install and power the active cooler. Figure 1.2 shows the Raspberry Pi 5 with the active cooler installed. The active cooler cools down the on-board SoC, RAM, and the Southgate chipset. With the active cooler and when the CPU is idle, the CPU temperature is around 40 degrees. The fan of the cooler operates automatically when the CPU temperature just exceeds 50 degrees Celsius.

Figure 1.2: Raspberry Pi 5 with active cooler.

The Raspberry Pi 5 operating system called **Bookworm** is available either on a pre-installed micro SD card or by downloading the system image onto a blank micro SD card. In this chapter, you will learn to install the operating system using both methods.

1.2 Using a pre-installed micro SD card

The pre-installed Raspberry Pi 5 operating system is available in various size micro SD cards. The author used the pre-installed 32 GB micro SD card supplied by Elektor. Additionally, the author used a 7-inch HDMI compatible monitor, a Raspberry Pi official keyboard, and a mouse. The author's hardware setup between the Raspberry Pi 5 and various devices is shown in Figure 1.3.

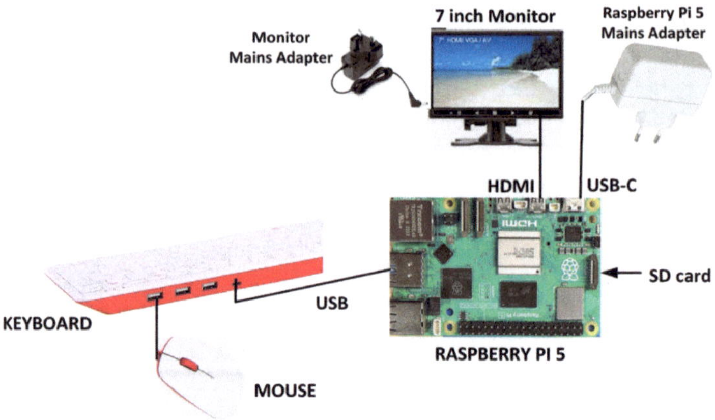

Figure 1.3: The authors' hardware setup.

The steps are as follows:

- Insert the pre-installed micro SD card into your Raspberry Pi 5.
- Connect all the devices as in Figure 1.3.
- Connect the Raspberry Pi 5 adapter to the mains (AC power).
- You should see the Raspberry Pi 5 booting the first time and asking you various questions to setup the device, such as the username, password, WiFi name and password, any updates if necessary, etc. etc. In this book, the username is set to **pi**.
- The Raspberry Pi 5 will boot in Desktop mode and will display the default screen. You can press Cntrl+Alt+F1 at any time to change to the Console mode.

1.3 Larger font in Console mode

It is probably hard to see the characters on a 7-inch monitor in Console mode. You can follow the steps below to increase the font size:

- Make sure you are in the Console mode
- Enter the following command:

  pi@raspberrypi: ~ $ **sudo dpkg-reconfigure console-setup**

- Select **UTF-8** in the **Package Configuration** screen (Figure 1.4)

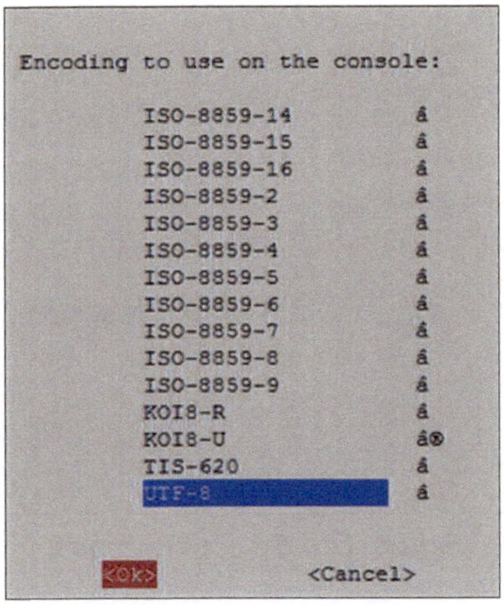

Figure 1.4: Select UTF-8.

- Select **Guess optimal character set** (Figure 1.5)

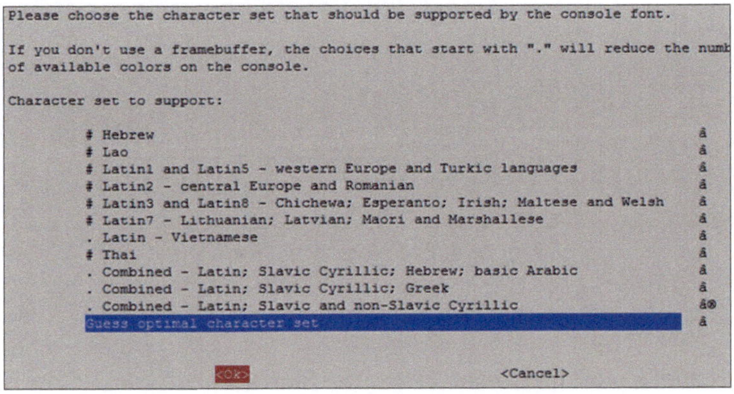

Figure 1.5: Select Guess optimal character set.

- Select **Terminus** (Figure 1.6)

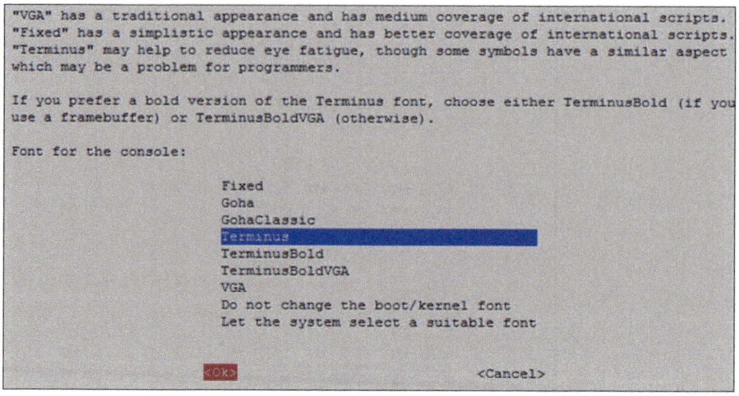

Figure 1.6: Select Terminus.

- Select font **16x32** (Figure 1.7)

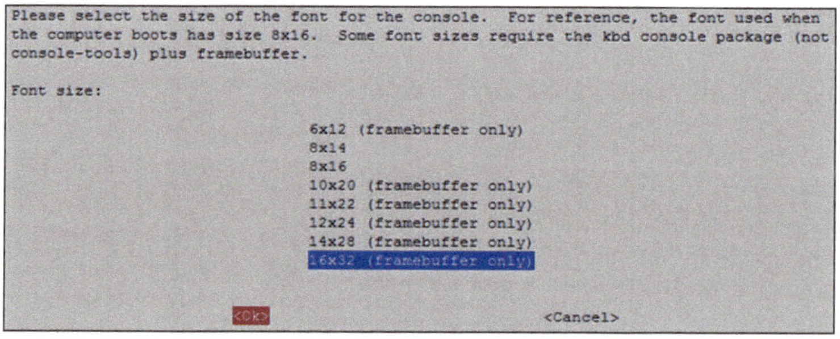

Figure 1.7: Select font 16×32.

1.4 Accessing your Raspberry Pi 5 Console from your PC – the Putty Program

In many applications, you may want to access your Raspberry Pi 5 from your PC. This requires enabling the SSH on your Raspberry Pi and then using a terminal emulation software on your PC. The steps to enable the SSH are as follows:

- Make sure you are in Console mode.
- Type: **sudo raspi-config**.
- Move down to **Interface Options**.
- Highlight **SSH** and press Enter (Figure 1.8).

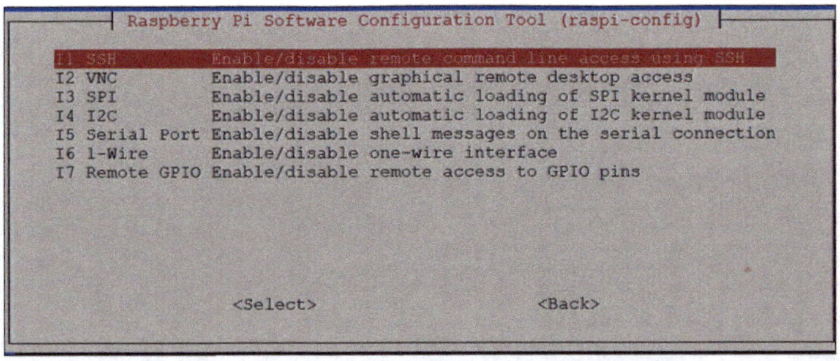

Figure 1.8: Highlight SSH.

- Click **Yes** to enable SSH.
- Click **OK**.
- Move down and click **Finish**.

You will now have to install a terminal emulation software on your PC. The one used by the authors is the popular 'Putty'. Download Putty from the following web site:

https://www.putty.org

- Putty is a standalone program and there is no need to install it. Simply double click to run it. You should see the Putty startup screen as in Figure 1.9.

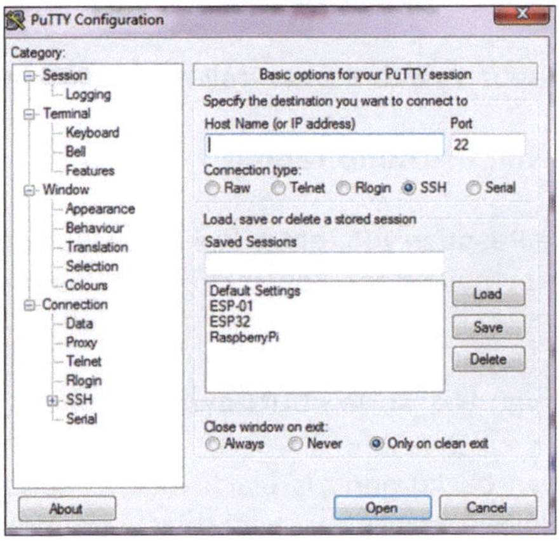

Figure 1.9: Putty startup screen.

- Make sure that the Connection type is SSH and enter the IP address of your Raspberry Pi 5. You can obtain the IP address by entering the command **ifconfig** in Console mode (Figure 1.10). In this example, the IP address was: **191.168.1.251** (see under **wlan0:**)

```
pi@raspberrypi:~ $ ifconfig
eth0: flags=4099<UP,BROADCAST,MULTICAST>  mtu 1500
        ether d8:3a:dd:77:b2:e2  txqueuelen 1000  (Ethernet)
        RX packets 0  bytes 0 (0.0 B)
        RX errors 0  dropped 0  overruns 0  frame 0
        TX packets 0  bytes 0 (0.0 B)
        TX errors 0  dropped 0 overruns 0  carrier 0  collisions 0
        device interrupt 107

lo: flags=73<UP,LOOPBACK,RUNNING>  mtu 65536
        inet 127.0.0.1  netmask 255.0.0.0
        inet6 ::1  prefixlen 128  scopeid 0x10<host>
        loop  txqueuelen 1000  (Local Loopback)
        RX packets 105  bytes 9175 (8.9 KiB)
        RX errors 0  dropped 0  overruns 0  frame 0
        TX packets 105  bytes 9175 (8.9 KiB)
        TX errors 0  dropped 0 overruns 0  carrier 0  collisions 0

wlan0: flags=4163<UP,BROADCAST,RUNNING,MULTICAST>  mtu 1500
        inet 192.168.1.251  netmask 255.255.255.0  broadcast 192.168.1.255
        inet6 2a00:23c7:868d:7b01:1562:5802:73c0:1ff6  prefixlen 64  scopeid 0x
<global>
        inet6 fe80::966a:a2d8:912:fa6b  prefixlen 64  scopeid 0x20<link>
```

Figure 1.10: Command ifconfig.

- Click **Open** in Putty after entering the IP address and selecting **SSH**.
- The first time you run Putty you may get a security message. Click **Yes** to accept this security alert.
- You will then be prompted to enter the Raspberry Pi 5 username and password. You can now enter all Console based commands through your PC.
- To change your password, enter the following command:

  pi@raspberrypi: ~ $ **passwd**

- To restart the Raspberry Pi enter the following command:

  pi@raspberrypi: ~ $ **sudo reboot**

- To shut down the Raspberry Pi, enter the following command. By the way, never shut down by pulling the power cable as this may result in the corruption or loss of files!

  pi@raspberrypi: ~ $ **sudo shutdown –h now**

By default, the **Putty** screen background is black with white foreground characters. The authors prefer to have white background with black foreground characters, with the character size set to 12 points bold. You should save your settings so that they are available next time you want to use the Putty. The steps to configure the Putty with these settings are given below:

- Restart Putty.
- Select **SSH** and enter the Raspberry Pi IP address.

- Click **Colours** under **Window**.
- Set the **Default Foreground** and **Default Bold Foreground** colours to black (Red:0, Green:0, Blue:0).
- Set the **Default Background** and **Default Bold Background** to white (Red:255, Green:255, Blue:255).
- Set the **Cursor Text** and **Cursor Colour** to black (Red:0, Green:0, Blue:0).
- Select **Appearance** under **Window** and click **Change** in **Font settings**. Set the font to **Bold 11**.
- Select **Session** and give a name to the session (e.g. MyZero) and click **Save**.
- Click **Open** to open the **Putty** session with the saved configuration.
- Next time you re-start the **Putty**, select the saved session and click **Load** followed by **Open** to start a session with the saved configuration.

1.5 Accessing the Desktop GUI from your PC

If you are using your Raspberry Pi 5 with a local keyboard, mouse, and display you can skip this section. If, on the other hand, you want to access your Desktop remotely over the network, you will find that SSH services cannot be used. The easiest and simplest way to access your Desktop remotely from a computer is by using the VNC (Virtual Network Connection) client and server. The VNC server runs on your Pi and the VNC client runs on your computer. It is recommended to use the **tightvncserver** on your Raspberry Pi 5. The steps are:

- Enter the following command:

  pi$raspberrypi:~ $ **sudo apt-get install tightvncserver**

- Run the **tightvncserver**:

  pi$raspberrypi:~ $ **tightvncserver**

You will be prompted to create a password for remotely accessing the Raspberry Pi desktop. You can also set-up an optional read-only password. The password should be entered every time you want to access the Desktop. Enter a password and remember it.

- Start the VNC server after reboot by the following command:

  pi$raspberrypi:~ $ **vncserver :1**

You can optionally specify screen pixel size and colour depth in bits as follows:

pi$raspberrypi:~ $ **vncserver :1 –geometry 1920x1080 –depth 24**

- You should now set up a VNC viewer on our laptop (or desktop) PC. There are many VNC clients available, but the recommended one which is compatible with **TightVNC** is the **TightVNC** for the PC which can be downloaded from the following link:

  https://www.tightvnc.com/download.php

- Download and install the **TightVNC** software for your PC. You will have to choose a password during the installation.

- Start the **TightVNC Viewer** on your PC and enter the Raspberry Pi IP address followed by :1. Click **Connect** to connect to your Raspberry Pi (Figure 1.11)

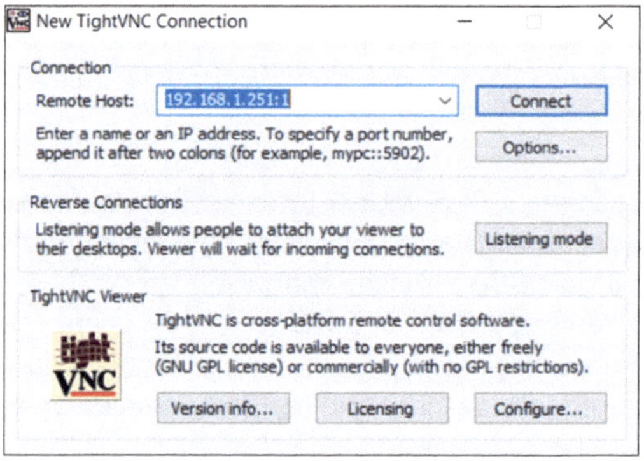

Figure 1.11: Connect to TightVNC Viewer.

- Enter the password you have chosen earlier. You should now see the Raspberry Pi 5 Desktop displayed on your PC screen (Figure 1.12)

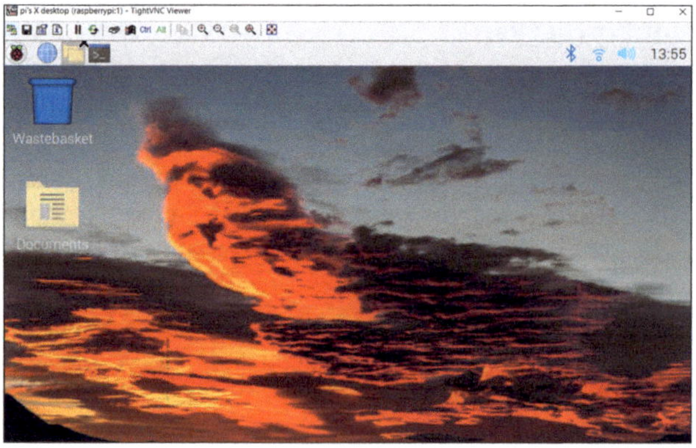

Figure 1.12: Raspberry Pi 5 Desktop.

- The VNC server is now running on your Raspberry Pi 5 and you have access to the Desktop GUI.

1.6 Assigning a Static IP Address to your Raspberry Pi 5

When you try to access your Raspberry Pi 5 remotely over your local network, it is possible that the IP address given by your Wi-Fi router changes from time to time. This is annoying as you have to find out the new IP address allocated to your Raspberry Pi. Without knowing the IP address, you cannot log in using the SSH or the VNC.

In this section, you will learn how to fix your IP address so that it does not change between reboots. The steps are as follows:

- Log in to your Raspberry Pi 5 via Putty.

- Check whether DHCP is active on your Raspberry Pi (it should normally be active):

  pi@raspberrypi:~ $ **sudo service dhcpcd status**

  If DHCP is not active, activate it by entering the following commands:

  pi@raspberrypi:~ $ **sudo service dhcpcd start**
  pi@raspberrypi:~ $ **sudo systemctl enable dhcpcd**

- Find the IP address currently allocated to you by entering command **ifconfig** or **hostname – I** (Figure 1.13). In this example the IP address was: 191.168.1.251. You can use this IP address as your fixed address since no other device on the network is currently using it.

```
pi@raspberrypi:~ $ hostname -I
192.168.1.251 2a00:23c7:868d:7b01:1562:5802:73c0:1ff6
pi@raspberrypi:~ $
```

Figure 1.13: Find the IP address using command hostname -I.

- Find the IP address of your router by entering the command **ip r** (Figure 1.14). In this example, the IP address was: 191.168.1.254.

```
pi@raspberrypi:~ $ ip r
default via 192.168.1.254 dev wlan0 proto dhcp src 192.168.1.251 metric 600
192.168.1.0/24 dev wlan0 proto kernel scope link src 192.168.1.251 metric 600
pi@raspberrypi:~ $
```

Figure 1.14: Find the IP address of your router.

- Find the IP address of your DNS by entering the following command (Figure 1.15). This is usually same as your router address:

  pi@raspberrypi:~ $ **grep "nameserver" /etc/resolv.conf**

```
pi@raspberrypi:~ $ grep "nameserver" /etc/resolv.conf
nameserver 192.168.1.254
nameserver fe80::4e1b:86ff:feb5:ba79%wlan0
pi@raspberrypi:~ $
```

Figure 1.15: Find the DNS address.

- Edit file **/etc/dhcpcd.conf** by entering the command:

  pi@raspberrypi:~ $ **nano /etc/dhcpcd.conf**

- Add the following lines to the bottom of the file (these will be different for your router). If these lines already exist, remove the comment character **#** at the beginning of the lines and change the lines as follows (you may notice that **eth0** for Ethernet is listed):

  interface wlan0
  static_routers=191.168.1.254
  static domain_name_servers=191.168.1.254
  static ip_address=191.168.1.251/24

- Save the file by entering **CNTRL + X** followed by **Y** and reboot your Raspberry Pi.

- In this example, the Raspberry Pi should reboot with the static IP address: 191.168.1.251.

1.7 Enabling Bluetooth

In this section, you will see how to enable the Bluetooth on your Raspberry Pi 5 so that it can communicate with other Bluetooth devices. The steps are given below:

- Enable the Bluetooth on your other device.

- Click on the Bluetooth icon on your Raspberry Pi 5 at the top right hand side and select **Make Discoverable.** You should see the Bluetooth icon flashing.

- Select raspberrypi in the Bluetooth menu on your other device.

- Accept the pairing request on your Raspberry Pi 5.

- You should now see the message **Connected Successfully** on your Raspberry Pi 5. and you can exchange files between your other device and the Raspberry Pi computer.

1.8 Connecting the Raspberry Pi 5 to a Wired Network

You may want to connect your Raspberry Pi 5 to a network through an Ethernet cable. The steps are as follows:

Step 1: Connect a network cable between your Raspberry Pi 5 and your WiFi router.

Step 2: Connect keyboard, mouse and monitor to your Raspberry Pi and power up as normal.

Step 3: Log in to the system by entering your username and password.

Step 4: Providing your network hub supports DHCP (nearly all network routers support DHCP), you will be connected automatically to the network and will be assigned a unique IP address within your network. Note that DHCP assigns IP addresses to newly connected devices.

Step 5: Check to find out the IP address assigned to your Raspberry Pi 5 by the network router. Enter command **ifconfig** as described earlier.

1.8.1 Unable to connect to a wired network

If you find that you are not assigned an IP address by the DHCP, the possible causes are:

- your network cable is faulty;
- the network hub does not support DHCP;
- your Raspberry Pi is not enabled to accept DHCP issued addresses. i.e. it may have been configured for fixed IP addresses.

In most cases it is very unlikely that the network cable is faulty. Also, most network hubs support the DHCP protocol. If you are having problems with the network, it is possible that your Raspberry Pi is not configured to accept DHCP issued addresses. The Raspberry Pi is normally configured to accept DHCP addresses but it is possible that you have changed the configuration somehow.

To resolve the wired network connectivity problem, follow the steps given below.

Step 1: find out whether or not your Raspberry Pi is configured for DHCP or fixed IP addresses. Enter the following command:

pi@raspberrypi ~$ **cat /etc/network/interfaces**

If your Raspberry Pi is configured to use the DHCP protocol (which is normally the default configuration), the word **dhcp** should appear at the end of the following line:

```
iface eth0 inet dhcp
```

If on the other hand your Raspberry Pi is configured to use static addresses then you should see the word **static** at the end of the following line:

```
iface eth0 inet static
```

Step 2: To use the DHCP protocol, edit file **interfaces** (e.g. using the **nano** text editor) and change the word **static** to **dhcp**. It is recommended to make a backup copy of the file interfaces before you change it:

```
pi@raspberrypi ~$ sudo cp /etc/network/interfaces /etc/network/int.bac
```

You should now restart your Raspberry Pi and an IP address will probably be assigned to your device.

Step 3: To use static addressing, make sure that the word **static** appears as shown above. If not, edit file **interfaces** and change **dhcp** to **static**

Step 4: You need to edit and add the required unique IP address, subnet mask and gateway addresses to file **interfaces** as in the following example (this example assumes that the required fixed IP address is 191.168.1.251, the subnet mask used in the network is 255.255.255.0, and the gateway address is 191.168.1.1):

```
iface eth0 inet static
address 191.168.1.251
netmask 255.255.255.0
gateway 191.168.1.1
```

Save the changes and exit the editor. If you are using the **nano** editor, exit by pressing Ctrl+X, and then enter Y to save the changes, and enter the filename to write to as **/etc/network/interfaces**.

Re-start your Raspberry Pi 5.

1.9 Installing the Raspberry Pi 5 Bookworm Operating System on a Blank microSD Card

If you have a pre-installed 'Bookworm' Raspberry Pi operating system on a microSD card then you can start using it as described earlier in this chapter. In this section, you will learn how to install the latest Bookworm operating system on a microSD card if you do not have a pre-installed card.

The steps are as follows:

- Insert a microSD card into your PC. You may need to use an SD card adapter.
- Go to the website: https://www.raspberrypi.com/software/
- Click to download the **Raspberry Pi Imager.** At the time of writing this book this file was called: **imager_1.7.5.exe**.
- Double click to start the imager program and click to install it.

- Click to **Finish** to run the imager.

- Click **Operating System** and select the operating system at the top of the list as: **Raspberry Pi OS (64-bit)**. See Figure 1.16.

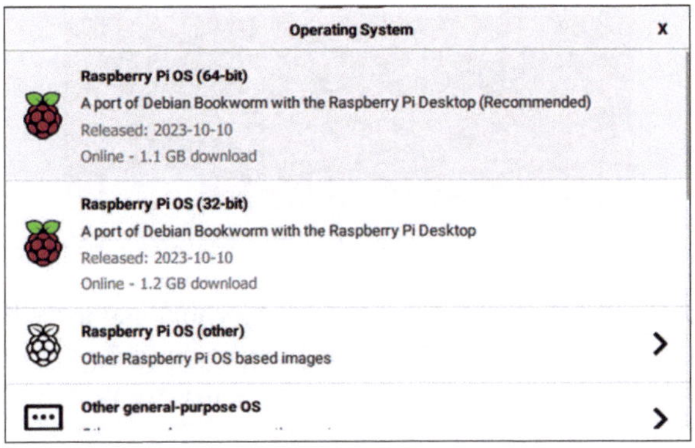

Figure 1.16: Select the operating system.

- Click **Storage** and select the SD card storage.

- Click to open the settings (gear shape).

- Click to enable SSH.

- Click to enable password authentication.

- Set username and password.

- Click to **Configure wireless LAN**.

- Click **Save**.

- Click **Write** to write the operating system to the microSD card.

- Wait until writing and verifying are finished (Figure 1.17).

- Remove the microSD card and insert into your Raspberry Pi 5.

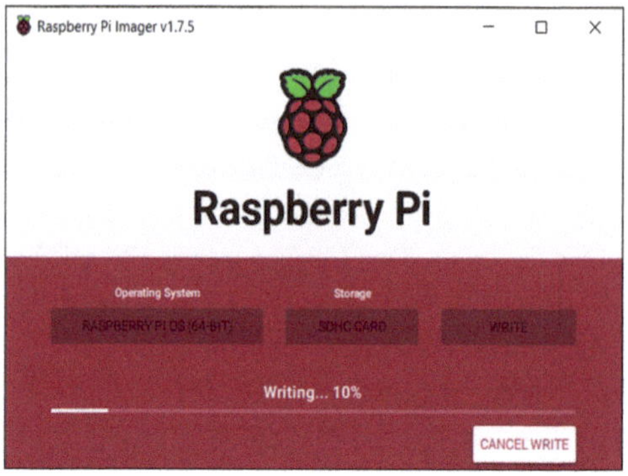

Figure 1.17: Writing to the micro SD card.

If you have a monitor and keyboard, you can log in to your raspberry Pi 5 directly and start using it. Otherwise, find the IP address of your Raspberry Pi 5 (e.g., from your router, or there are many apps for smartphones, such as **who's on my wifi**, showing all the devices connected to your router with their IP addresses). Then log in to your Raspberry Pi 5 and start using it.

Chapter 2 • Using a Text editor, Creating and Running a Python Program

A text editor is used to create or modify the contents of a text file. There are many text editors available for the Linux operating system. Some popular ones are nano, vim, vi, and many more. In this Chapter, you will be learning how to use the popular text editor called **nano**. Additionally, you will be learning the different methods of creating and running a Python program.

2.1 The nano Text Editor

This is probably the most commonly used text editor. Start the **nano** text editor by entering the word **nano**, followed by the filename you wish to create or modify. An example is given below, where a new file called **first.txt** is created:

pi@raspberrypi: ~ $ **nano first.txt**

You should see the editor screen as in Figure 2.1. The name of the file to be created or modified is displayed at the top middle part of the screen. The message **New File** at the bottom of the screen shows that this is a newly created file. The shortcuts at the bottom of the screen are there to perform various editing functions. These shortcuts are accessed by pressing the Ctrl key together with another key. Some of the useful shortcuts are described below:

Ctrl+W: Search for a word.
Ctrl+V: Move to next page.
Ctrl+Y: Move to previous page.
Ctrl+K: Cut the current row of txt.
Ctrl+R: Read file.
Ctrl+U: Paste the text you previously cut.
Ctrl+J: Justify.
Ctrl+\: Search and replace text.
Ctrl+C: Display current column and row position.
Ctrl+G: Get detailed help on using nano.
Ctrl+-: Go to specified line and column position.
Ctrl+O: Save (write out) the file currently open.
Ctrl+X: Exit nano.

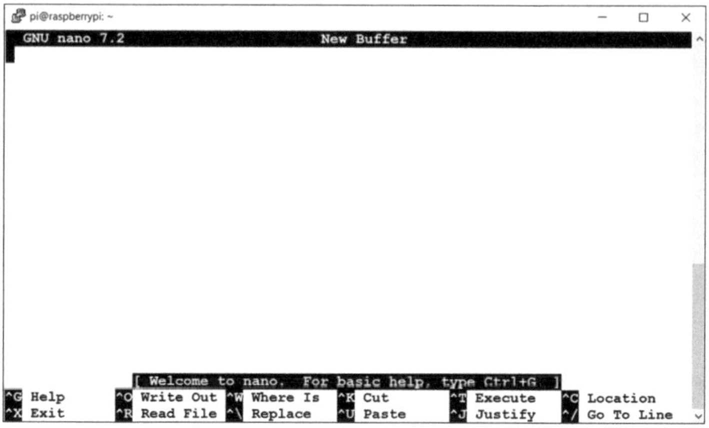

Figure 2.1: nano text editor screen.

Now, type the following text as shown in Figure 2.2:

> nano is a simple and yet powerful text editor.
> This simple text example demonstrates how to use nano.
> This is the last line of the example.

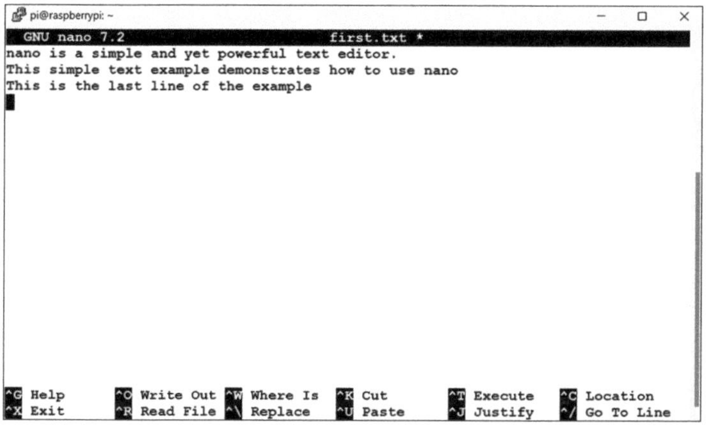

Figure 2.2: Sample text.

The use of **nano** is now demonstrated with the following steps.

Step 1: Go the beginning of the file by moving the cursor up and left.

Step 2: Look for word **simple** by pressing **Ctrl+W** and then typing **simple** in the window opened at the bottom left hand corner of the screen. Press the Enter key. The cursor will be positioned on the word **simple** (see Figure 2.3).

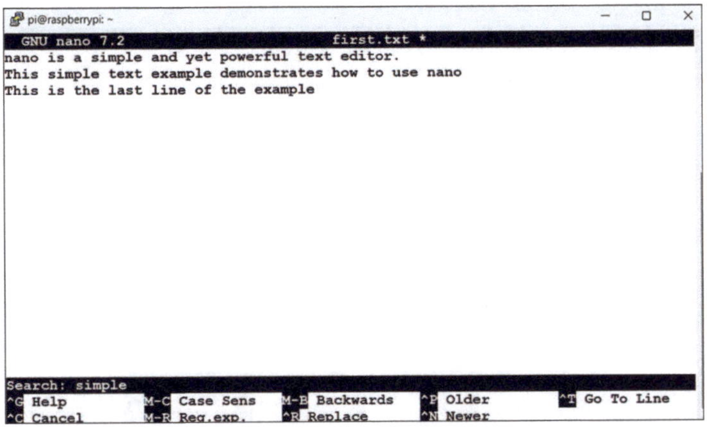

Figure 2.3: Searching the word 'simple'.

Step 3: Cut the first line by placing the cursor anywhere on the line and then pressing **Ctrl+K**. The first line will disappear as in Figure 2.4.

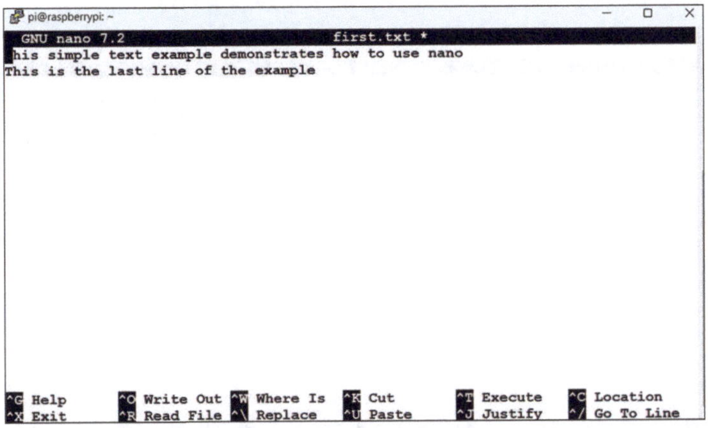

Figure 2.4: Cutting the first line.

Step 4: Paste the line cut after the second line. Place the cursor on the third line and press **Ctrl+U** (see Figure 2.5).

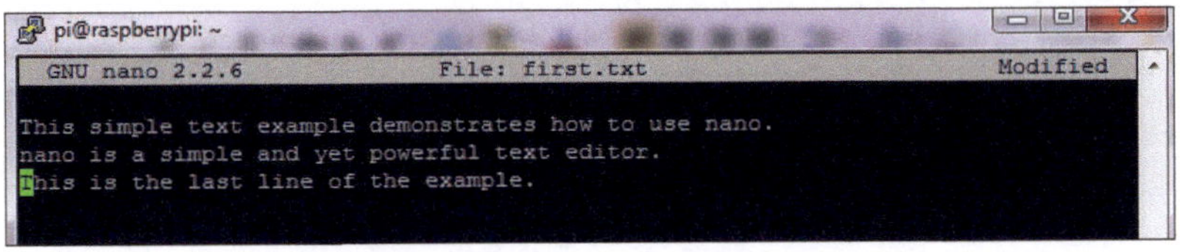

Figure 2.5: Paste the line cut previously.

Step 5: Place cursor at the beginning of word **simple** on the first line. Enter **Ctrl+C**. The row and column positions of this word will be displayed at the bottom of the screen (Figure 2.6).

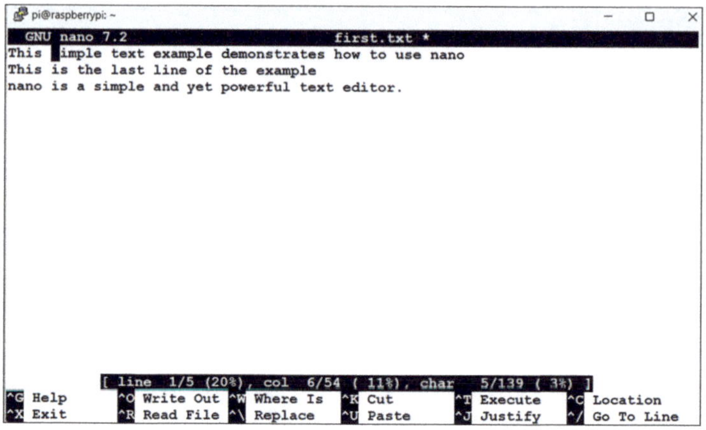

Figure 2.6: Displaying row and column position of a word.

Step 6: Press **Ctrl+G** to display help page as in Figure 2.7. Notice that the display is many pages long and you can jump to the next pages by pressing **Ctrl+Y** or to the previous pages by pressing **Ctrl+V**. Press **Ctrl+X** to exit the help page.

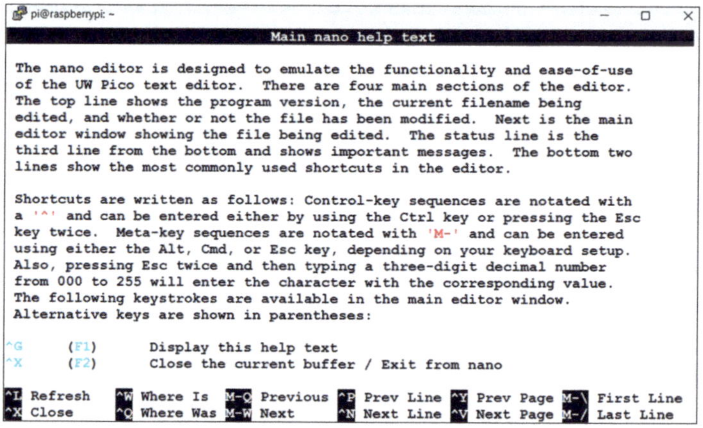

Figure 2.7: Displaying the Help page.

Step 7: Press **Ctrl+-** and enter line and column numbers as 2 and 5, followed by the Enter key, to move cursor to line 2, column 5 (Figure 2.8).

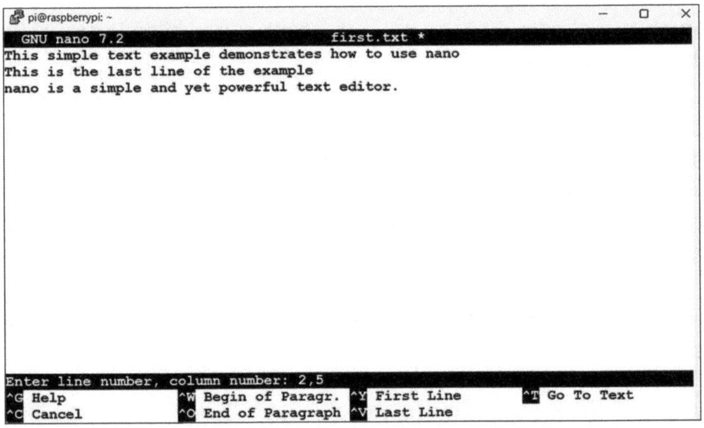

Figure 2.8: Move cursor to specified line and column.

Step 8: Replace word **example** with word **file**. Press **Ctrl+** and type the first word as **example** (see Figure 2.9). Press Enter and then type the replacement word as **file**. Press Enter and accept the change by typing y.

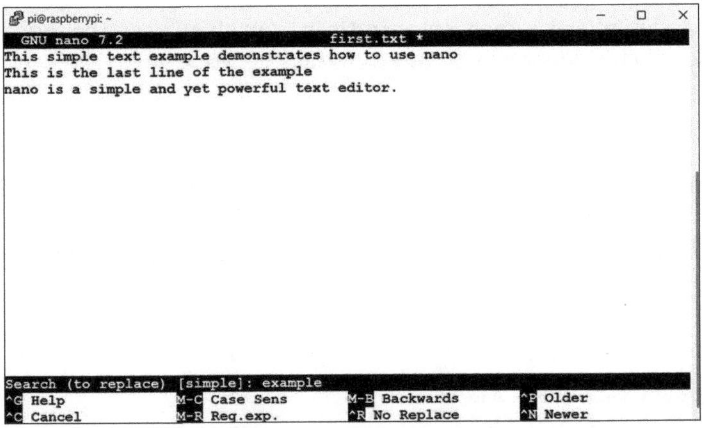

Figure 2.9: Replacing text.

Step 9: Save the changes. Press **Ctrl+X** to exit the file. Type **Y** to accept the saving, then enter the filename to be written to, or simply press Enter to write to the existing file (**first.txt** in this example). The file will be saved in your current working directory.

Step 10: Display the contents of the file:

```
pi@raspberrypi: ~ $ cat first.txt
This simple text file demonstrates how to use nano.
This is the last line of the example.
Nano is a simple and yet powerful text editor
pi@raspberrypi: ~ $
```

In summary, **nano** is a simple and yet powerful text editor allowing us to create new text files or to modify existing files.

2.2 Creating and Running a Python Program

You're now ready to start programming your Raspberry Pi 5 using the Python language. It is worthwhile to look at the creation and running of a simple Python program on your Raspberry Pi 5 computer. In this chapter, the message **Hello From Raspberry Pi 5** will be displayed on your PC screen.

As described below, there are 3 methods that you can create and run Python programs on your Raspberry Pi 5.

2.2.1 Method 1 — Interactively from command prompt in Console mode

In this method, you will login to your Raspberry Pi 5 using the SSH and then create and run the Python program interactively. This method is excellent for small programs. The steps are as follows:

- Log in to the Raspberry Pi 5 using SSH.
- At the command prompt, enter **python**. You should see the Python command mode which is identified by three characters **>>>**
- Type the program:

    print ("Hello From Raspberry Pi 5")

- The text will be displayed interactively on the screen as shown in Figure 2.10. Note that at the time of writing this book the Python version was: 3.11.2.

```
pi@raspberrypi:~ $ python
Python 3.11.2 (main, Mar 13 2023, 12:18:29) [GCC 12.2.0] on linux
Type "help", "copyright", "credits" or "license" for more information.
>>> print("Hello from Raspberry Pi 5")
Hello from Raspberry Pi 5
>>> 
```

Figure 2.10 Running a Python program interactively

- Type **Cntrl+z** to exit from the program.

In this method, you will log in to your Raspberry Pi 5 using the SSH as before and then create a Python file. A Python file is simply a text file with the extension **.py**. You can use a text editor, e.g. the **nano** text editor to create your file. In this example, a file called **hello.py** is created using the **nano** text editor. Figure 2.11 shows the contents of file **hello.py**. This figure also shows how to run the file under Python. Notice that the program is run by entering the command:

>>> python hello.py

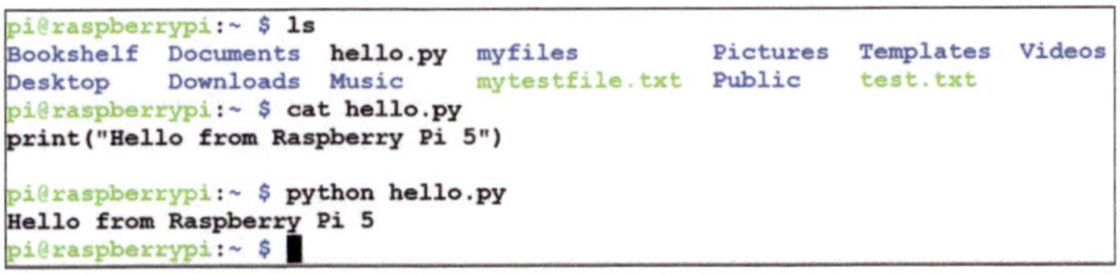

Figure 2.11: Creating and running a Python file.

In this method, you can login to your Raspberry Pi 5 using either a directly connected monitor and keyboard through the mini HDMI port, or if you don't have a monitor, you can login to the Desktop using the VNC as described earlier, and then create and run your Python programs in GUI mode using the Thonny IDE. It is worthwhile at this stage to learn the basics of using the Thonny IDE.

The 'Thonny' IDE

Start the Thonny IDE from the Desktop GUI under the **Programming** menu. Figure 2.12 shows the Thonny start up menu.

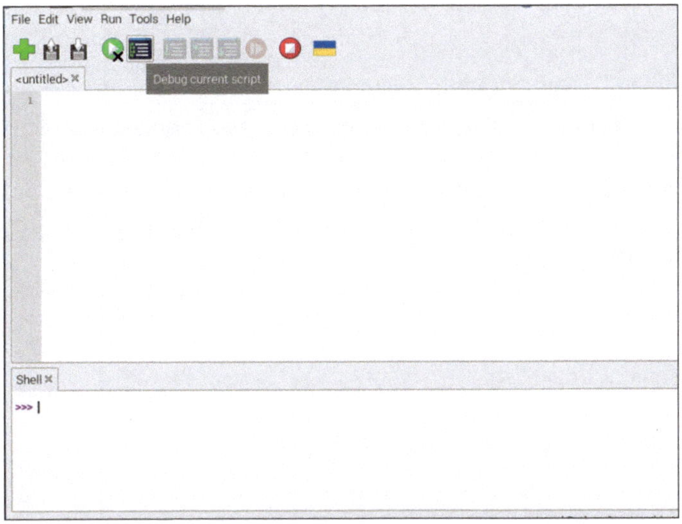

Figure 2.12: Thonny IDE startup menu.

The screen consists of two parts: the upper part is where you write your programs, and the lower part is the shell where small interactive program codes can be written. This part is mainly used for testing small program codes.

In the upper part contains the following menu items:

File: Open, close, or save a file.
Edit: Cut, paste, indent, select, toggle comment, uncomment, find & replace, clear shell.
View: Exception, files, help, notes, shell, stack, variables, font size, plotter.
Run: Configure interpreter, run script, debug script, stop/start, interrupt execution.

Tools: Manage packages, open program folder, open data folder, and manage plug-ins.
Help: Help contents, version history, report problems, about Thonny.

The Thonny IDE must be configured before it is used to write and upload programs to your Raspberry Pi 5. Click the bottom right corner of the screen to select your processor type and select **Local Python 3**. You are now ready to write your program. The steps are:

- Type the following code to the upper part of the screen:

```
print("Hello from Raspberry Pi 5")
```

- Click **File** → **Save** and save with the name hello.py (Figure 2.13)

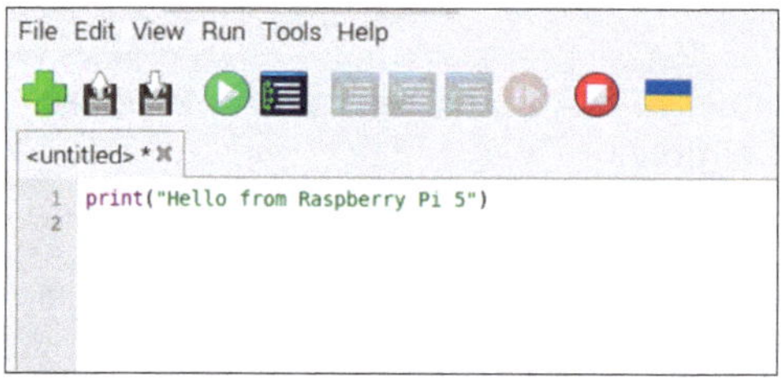

Figure 2.13: Type your program and then save it.

- Click the **Run** icon (green menu button at the top) to run the program. The output of the program will be displayed at the bottom of the screen as shown in Figure 2.14.

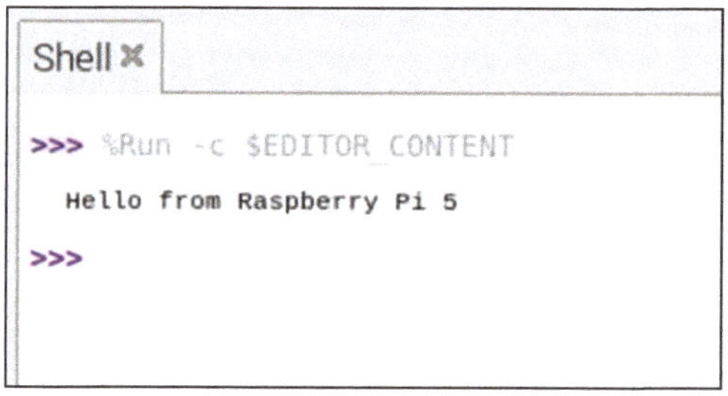

Figure 2.14: Run the program.

You can run small programs in interactive mode by entering them at the lower part of the screen under **Shell**. The results will be displayed under **Shell** immediately.

2.3 Which method?

The choice of a method depends upon the size and complexity of a program. Small programs can be run interactively in Shell under Thonny or directly in the Console mode (after typing **python**) without creating a program file. Larger programs can be created as Python files using the **nano** text editor and then they can then run either in the Console mode (after typing **python** followed by the filename with extension **.py**), or in Desktop GUI mode under the **Thonny** IDE. Running under the **Thonny** IDE has the advantage that justification of the code is corrected automatically by the IDE as you write the code. In this book, the Thonny IDE is used for small programs and the **nano** text editor is used for larger programs to create the program files.

Chapter 3 • Amateur Radio Programs – Software-Only

3.1 Overview

This chapter presents simple electronic engineering based Python programs for your Raspberry Pi 5 with a particular interest to radio amateurs designing and analyzing elementary RF circuitry. The programs discussed are software-only, meaning they do not use any external hardware to produce results.

3.2 4-Band Resistor Color Code Identifier

In this project, the user enters the three colors of a 4-band resistor and the program calculates and displays the value of the resistor in Ohms. The tolerance of the resistor is not displayed.

Resistor values are identified by the following color codes:

Black:	0
Brown:	1
Red:	2
Orange:	3
Yellow:	4
Green:	5
Blue:	6
Violet:	7
Grey:	8
White:	9

The first two color determine the first two digits of the value while the last color determines the multiplier. For example, **red red brown** corresponds to 22 × 10 = 220 ohms (Figure 3.1).

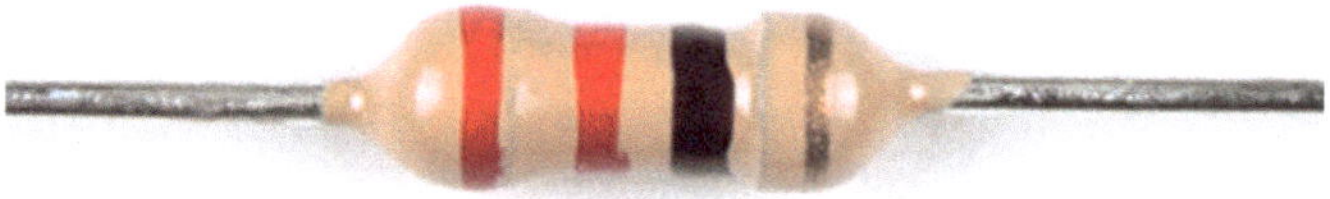

Figure 3.1: A 220-ohm resistor,

Program listing: Figure 3.2 shows the program listing (program: **resistor.py**). At the beginning of the program a list called **color** is created which stores the valid resistor colors. Then a heading is displayed and a **while** loop is created which runs as long as string variable **yn** is equal to y. Inside the loop the program reads the three colors from the keyboard using functions **input** and stores as strings in variables **FirstColour**, **SecondColour** and **ThirdColour**. These strings are then converted into lower case so that they are compatible with the values listed in the list box. The index values of these colors in the list are then found using function calls of the form **colours.index**. Remember that the index values start from 0. As an example if the user entered red then the corresponding index value will be 2. The resistor value is then calculated by multiplying the first color number by 10 and adding to the second color number. The result is then multiplied by the power of 10 of the third color index. The final result is displayed on the screen. The program then asks whether or not the user wants to continue. If the answer is **y** then the program returns to the beginning, otherwise the program is terminated.

```
#=====================================================
#                RESISTOR COLOUR CODES
#                ---------------------
#
# The user enters the three colours of a resistor
# and the program calculates and displays the value
# of the resistor in Ohms
#
# Program: resistor.py
# Date   : February, 2024
# Author : Dogan Ibrahim
#=====================================================
colours = ['black','brown','red','orange','yellow','green',\
'blue','violet','grey','white']

print("RESISTOR VALUE CALCULATOR")
print("=========================")
yn = "y"

while yn == 'y':
  FirstColour = input("Enter First Colour: ")
  SecondColour = input("Enter Second Colour: ")
  ThirdColour = input("Enter Third Colour: ")
#
# Convert to lower case
#
  FirstColour = FirstColour.lower()
  SecondColour = SecondColour.lower()
  ThirdColour = ThirdColour.lower()
#
# Find the values of colours
#
  FirstValue = colours.index(FirstColour)
  SecondValue = colours.index(SecondColour)
  ThirdValue = colours.index(ThirdColour)
#
# Now calculate the value of the resistor
#
  Resistor = 10 * FirstValue  + SecondValue
  Resistor = Resistor * (10 ** ThirdValue)
  print("Resistance = %d Ohms" % (Resistor))
```

```
#
# Ask for more
#
  yn = input("\nDo you want to continue?: ")
  yn = yn.lower()
```

Figure 3.2: Program listing.

Figure 3.3 shows a typical run of the program where 220-ohm and 2200-ohm resistors are identified. The program was run in Console mode with the following command:

pi@raspberrypi: ~ $ **python resistor.py**

```
pi@raspberrypi:~ $ python resistor.py
RESISTOR VALUE CALCULATOR
=========================
Enter First Colour: red
Enter Second Colour: red
Enter Third Colour: brown
Resistance = 220 Ohms

Do you want to continue?: y
Enter First Colour: red
Enter Second Colour: red
Enter Third Colour: red
Resistance = 2200 Ohms

Do you want to continue?: n
pi@raspberrypi:~ $
```

Figure 3.3: Typical run of the program.

3.3 4-Band Resistor Color Code Identifier Including Very Small Resistors

The program given in Figure 3.2 can calculate the values of resistors with a value greater than or equal to 10 ohms. For smaller resistors, gold and silver colors are used as the third color. Gold divides the result by 10 and silver divides by 100. For example, **red red silver** corresponds to 0.22 ohms (Figure 3.4). This program calculates the values of all types of resistors identified by 4 color bands.

Figure 3.4: A 0.22-ohm resistor.

Program listing: Figure 3.5 shows the program listing (program: **resistor2.py**). The colors gold and silver are added to the list. The indexes of gold and silver in the list are 10 and 11, respectively. Most of the program is same as in Figure 3.2, except that if the third color is gold then the result is divided by 10. Similarly, if the third color is silver then the result is divided by 100. Notice that if the resistor value is less than 10 ohms then it is displayed in floating point format with two digits after the decimal point by using the print formatting parameter %3.2f.

```
#==================================================
#              RESISTOR COLOUR CODES
#              ---------------------
#
# The user enters the three colours of a resistor
# and the program calculates and displays the value
# of the resistor in Ohms. The program identifies all
# types of resistors with 4 colour bands
#
# Program: resistor2.py
# Date   : February, 2024
# Author : Dogan Ibrahim
#====================================================
colours = ['black','brown','red','orange','yellow','green',\
'blue','violet','grey','white', 'gold', 'silver']

print("RESISTOR VALUE CALCULATOR")
print("=========================")
yn = "y"

while yn == 'y':
  FirstColour = input("Enter First Colour: ")
  SecondColour = input("Enter Second Colour: ")
  ThirdColour = input("Enter Third Colour: ")
#
# Convert to lower case
#
  FirstColour = FirstColour.lower()
  SecondColour = SecondColour.lower()
  ThirdColour = ThirdColour.lower()
#
# Find the values of colours
#
  FirstValue = colours.index(FirstColour)
  SecondValue = colours.index(SecondColour)
  ThirdValue = colours.index(ThirdColour)
#
# Now calculate the value of the resistor
#
  Resistor = 10 * FirstValue  + SecondValue
  if ThirdValue == 10:
```

```
    Resistor = Resistor / 10.0
    print("Resistance = %3.2f Ohms" % (Resistor))
  elif ThirdValue == 11:
    Resistor = Resistor/100.0
    print("Resistance = %3.2f Ohms" % (Resistor))
  else:
    Resistor = Resistor * (10 ** ThirdValue)
    print("Resistance = %d Ohms" % (Resistor))
#
# Ask for more
#
  yn = input("\nDo you want to continue?: ")
  yn = yn.lower()
```

Figure 3.5: Program listing.

Figure 3.6 shows a typical run of the program. The program was run in Console mode with the following command:

pi@raspberrypi: ~ $ **python resistor2.py**

```
pi@raspberrypi:~ $ python resistor2.py
RESISTOR VALUE CALCULATOR
=========================
Enter First Colour: red
Enter Second Colour: red
Enter Third Colour: gold
Resistance = 2.20 Ohms

Do you want to continue?: y
Enter First Colour: red
Enter Second Colour: red
Enter Third Colour: silver
Resistance = 0.22 Ohms

Do you want to continue?: n
pi@raspberrypi:~ $ █
```

Figure 3.6: Typical run of the program.

3.4 Series or Parallel Resistors

This program calculates the total resistance of a number of series- or parallel-connected resistors. The user specifies whether the connection is in series or in parallel. Additionally, the number of resistors used is also specified at the beginning of the program.

When a number of resistors are in series, the resultant resistance is the sum of the resistance of each resistor. When the resistors are in parallel, the reciprocal of the resultant resistance equals the sum of the reciprocal resistances of each resistor.

Program Listing: Figure 3.7 shows the program listing (program: **serpal.py**). At the beginning of the program a heading is displayed and the program enters into a **while** loop. Inside this loop, the user is prompted to enter the number of resistors in the circuit and whether they are connected in series or in parallel. Function **str** converts a number into its equivalent string. e.g., number 5 is converted into string "5". If the connection is serial (mode equals to **'s'**) then the value of each resistor is accepted from the keyboard and the resultant is calculated and displayed on the screen. If on the other hand the connection is parallel (mode is equals to **'p'**) then again the value of each resistor is accepted from the keyboard and the reciprocal of the number is added to the total. When all the resistor values are entered, the resultant resistance is displayed on the screen.

```
#====================================================
#         RESISTORS IN SERIES OR PARALLEL
#         ------------------------------
#
# This program calculates the total resistance of
# serial or parallel connected resistors
#
# Program: serpal.py
# Date   : February, 2024
# Author : Dogan Ibrahim
#====================================================
print("RESISTORS IN SERIES OR PARALLEL")
print("===============================")
yn = "y"

while yn == 'y':
  N = int(input("\nHow many resistors are there?: "))
  mode = input("Are the resistors series (s) or parallel (p)?: ")
  mode = mode.lower()
#
# Read the resistor values and calculate the total
#
  resistor = 0.0

  if mode == 's':
    for n in range(0,N):
       s = "Enter resistor " + str(n+1) + " value in Ohms: "
       r = int(input(s))
       resistor = resistor + r
    print("Total resistance = %d Ohms" %(resistor))

  elif mode == 'p':
    for n in range(0,N):
      s = "Enter resistor " + str(n+1) + " value in Ohms: "
      r = float(input(s))
      resistor = resistor + 1 / r
    print("Total resistance = %.2f Ohms" %(1 / resistor))
#
```

```
# Check if the user wants to exit
#
  yn = input("\nDo you want to continue?: ")
  yn = yn.lower()
```

Figure 3.7: Program listing.

Figure 3.8 shows a typical run of the program.

```
pi@raspberrypi:~ $ python serpal.py
RESISTORS IN SERIES OR PARALLEL
===============================

How many resistors are there?: 2
Are the resistors series (s) or parallel (p)?: s
Enter resistor 1 value in Ohms: 250
Enter resistor 2 value in Ohms: 100
Total resistance = 350 Ohms

Do you want to continue?: y

How many resistors are there?: 2
Are the resistors series (s) or parallel (p)?: p
Enter resistor 1 value in Ohms: 100
Enter resistor 2 value in Ohms: 100
Total resistance = 50.00 Ohms

Do you want to continue?: n
pi@raspberrypi:~ $
```

Figure 3.8: Typical run of the program.

3.5 Capacitor Identification

The marking of capacitor values depends on the type of capacitor used.

Electrolytic capacitors

As shown in Figure 3.9, the capacitance, maximum working voltage, and the polarity of these capacitors are marked on the body of the capacitor. In the image, the capacitor is 1000 µF and its maximum working voltage is16 V.

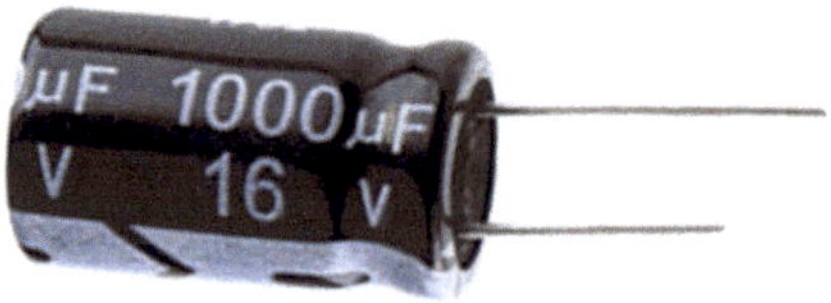

Figure 3.9: Electrolytic capacitor marking.

Ceramic capacitors

These capacitors have 3-digit numeric markings on them. First two digits are the capacitance values and the third digit is the multiplier as follows:

- **0**: multiply by **1**
- **1**: multiply by **10**
- **2**: multiply by **100**
- **3**: multiply by **1000**
- **4**: multiply by **10,000**
- **5**: multiply by **100,000**
- **6**: multiply by **1,000,000**

For example, the capacitor with marking 100 is 10 × 1 = 10 pF (Figure 3.10). Similarly,

101 is 10 × 10 = 100 pF
102 is 10 × 100 = 1000 pF
103 is 10 × 1000 = 10,000 pF (10 nF)

and so on.

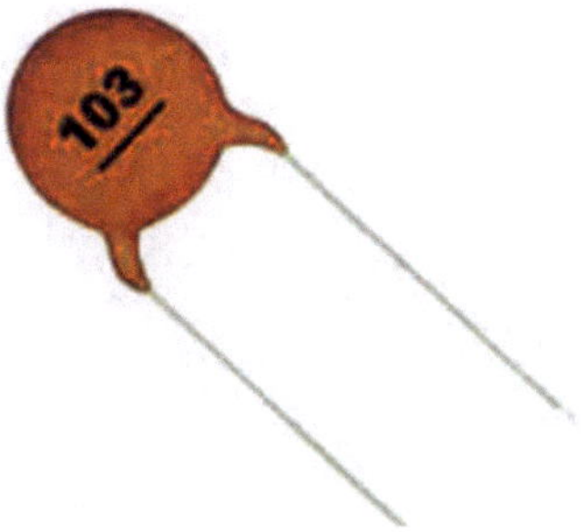

Figure 3.10: A 10,000-pF (10-nF) ceramic capacitor.

Program listing: Figure 3.11 shows the program listing (program: **capacitor.py**). At the beginning of the program, a heading is displayed and the program enters into a **while** loop. Inside this loop the user is prompted to enter the marking on the capacitor (e.g., 102). The program then calculates the value of the capacitor and displays on the screen.

```
#=========================================================
#                CAPACITOR IDENTIFICATION
#                ------------------------
#
# The user enters the marking on the capacitor (e.g. 102)
# and the program displays the capacitance in pF
#
# Program: capacitor.py
# Date   : February, 2024
# Author : Dogan Ibrahim
#=========================================================
print("CAPACITOR VALUE CALCULATOR")
print("==========================")
yn = "y"
```

```
while yn == 'y':
    Marking = input("Enter the marking on the capacitor (e.g. 102): ")
#
# Find the value and multiplier
#
    Multiplier = int(Marking[2])
    Value = int(Marking[0:2])
    Cap = Value *  10 ** Multiplier
    print("Capacitance = %d pF" % (Cap))
#
# Ask for more
#
    yn = input("\nDo you want to continue?: ")
    yn = yn.lower()
```

Figure 3.11: Program listing.

Figure 3.12 shows a typical run of the program.

```
pi@raspberrypi:~ $ python capacitor.py
CAPACITOR VALUE CALCULATOR
==========================
Enter the marking on the capacitor (e.g. 102): 102
Capacitance = 1000 pF

Do you want to continue?: y
Enter the marking on the capacitor (e.g. 102): 130
Capacitance = 13 pF

Do you want to continue?: y
Enter the marking on the capacitor (e.g. 102): 132
Capacitance = 1300 pF

Do you want to continue?: y
Enter the marking on the capacitor (e.g. 102): 133
Capacitance = 13000 pF

Do you want to continue?: y
Enter the marking on the capacitor (e.g. 102): 134
Capacitance = 130000 pF

Do you want to continue?: n
```

Figure 3.12: Typical run of the program.

3.6 Capacitors in Series or in Parallel

This program calculates the total capacitance of a number of series- or parallel-connected capacitors. The user specifies whether the connection is in series or in parallel. Additionally, the number of capacitors used is also specified at the beginning of the program.

When a number of capacitors are in parallel then the resultant capacitance is the sum of the capacitance of each capacitor. When the capacitors are in series then the reciprocal of the resultant capacitance is equal to the sum of the reciprocal capacitances of each capacitor.

Program Listing: Figure 3.13 shows the program listing (program: **capserpal.py**). At the beginning of the program a heading is displayed and the program enters into a **while** loop. Inside this loop, the user is prompted to enter the number of capacitors in the circuit and whether they are connected in series or in parallel. Function **str** converts a number into its equivalent string. e.g., number 5 is converted into string "5". If the connection is serial (mode equals to **'s'**) then the value of each capacitor is accepted from the keyboard and the resultant is calculated and displayed on the screen. If on the other hand the connection is parallel (mode is equals to **'p'**) then again the value of each capacitor is accepted from the keyboard and the reciprocal of the number is added to the total. When all the capacitor values are entered, the resultant capacitance is displayed on the screen. Note that the entered capacitors must all have the same units.

```
#=========================================================
#          CAPACITORS IN SERIES OR PARALLEL
#          -------------------------------
#
# This program calculates the total capacitance of
# serial or parallel connected capacitors
#
# Program: capserpal.py
# Date   : February, 2024
# Author : Dogan Ibrahim
#=========================================================
print("CAPACITORS IN SERIES OR PARALLEL")
print("================================")
print()
print("Use same units for all capacitors")
yn = "y"

while yn == 'y':
  N = int(input("\nHow many capacitors are there?: "))
  mode = input("Are the capacitors series (s) or parallel (p)?: ")
  mode = mode.lower()
#
# Read the capacitor values and calculate the total
#
  capacitor = 0.0

  if mode == 'p':
    for n in range(0,N):
       s = "Enter capacitor " + str(n+1) + ": "
       r = int(input(s))
       capacitor = capacitor + r
    print("Total capacitance = %d" %(capacitor))

  elif mode == 's':
```

```
    for n in range(0,N):
      s = "Enter capacitor " + str(n+1) + ": "
      r = float(input(s))
      capacitor = capacitor + 1 / r
    print("Total capacitance = %.2f" %(1 / capacitor))
#
# Check if the user wants to exit
#
  yn = input("\nDo you want to continue?: ")
  yn = yn.lower()
```

Figure 3.13: Program listing.

Figure 3.14 shows a typical run of the program.

```
pi@raspberrypi:~ $ python capserpal.py
CAPACITORS IN SERIES OR PARALLEL
================================

Use same units for all capacitors

How many capacitors are there?: 2
Are the capacitors series (s) or parallel (p)?: p
Enter capacitor 1: 20
Enter capacitor 2: 15
Total capacitance = 35

Do you want to continue?: y

How many capacitors are there?: 2
Are the capacitors series (s) or parallel (p)?: s
Enter capacitor 1: 20
Enter capacitor 2: 20
Total capacitance = 10.00

Do you want to continue?: n
pi@raspberrypi:~ $
```

Figure 3.14: Typical run of the program.

3.7 Resistive Potential Divider

Resistive potential divider circuits typically consist of two or more resistors. These circuits are used to lower a voltage to a desired value. Figure 3.15 shows a typical resistive potential divider circuit. Here, V_{in} and V_o are the input and output voltages, respectively. R1 and R2 is the resistor pair used to lower the voltage from V_{in} to V_o. A large number of resistor pairs can be used to get the desired output voltage. Choosing high-value resistors draws little current from the circuit, and choosing low-value resistors draws larger currents. In this design, the user specifies V_{in}, V_o, and R2. The program calculates the required R1 value to lower the voltage to the desired level. Additionally, the program displays the output voltage with the chosen physical resistors.

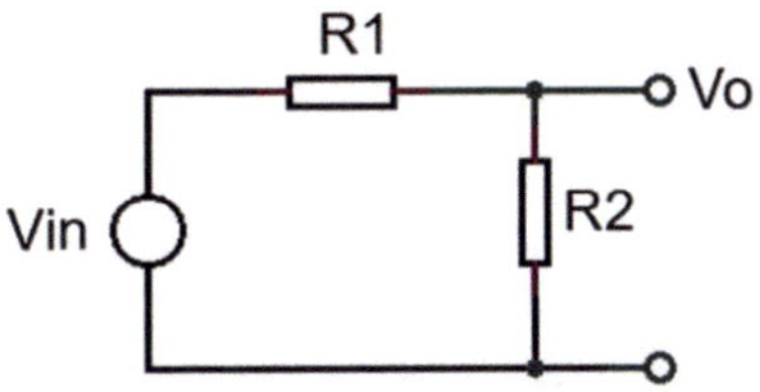

Figure 3.15: Resistive potential divider circuit.

The output voltage is given by:

$$V_o = V_{in}\ R2 / (R1 + R2)$$

R1 is then given by:

$$R1 = (V_{in} - V_o)\ R2/ V_o$$

The above formula is used to calculate the required value of R1, given V_{in}, V_o, and R2.

Program listing: Figure 3.16 shows the program listing (program: **divider.py**). At the beginning of the program, a heading is displayed. The program then reads V_{in}, V_o, and R2 from the keyboard. The program calculates R1 and displays R1 and R2. The user is then asked to enter a chosen physical value for R1. With the chosen value of R1, the program displays V_{in}, V_o, R1, and R2 and asks the user whether or not the result is acceptable. If the answer to this question is **y** then the program terminates. If on the other hand the answer is **n** then the user is given the option of trying again.

```
#=========================================================
#              RESISTIVE POTENTIAL DIVIDER
#              ---------------------------
#
# This is a resistive potential divider circuit program.
# The program calculates the resistance values that will
# lower the input voltage to the desired value
#
# Program: divider.py
# Date   : February, 2024
# Author : Dogan Ibrahim
#==========================================================
print("RESISTIVE POTENTIAL DIVIDER")
print("===========================")
R1flag = 1
R2flag = 0

while R1flag == 1:
  Vin = float(input("\nInput voltage (Volts): "))
  Vo = float(input("Desired output voltage (Volts): "))
```

```
    R2 = float(input("Enter R2 (in Ohms): "))
#
# Calculate R1
#
    R1 = R2 * (Vin - Vo) / Vo
    print("\nR1 = %3.2f Ohms R2 = %3.2f Ohms" %(R1, R2))
#
# Read chosen physical R1 and display actual Vo
#
    NewR1 = float(input("\nEnter chosen R1 (Ohms): "))

#
# Display and print the output voltage with chosen R1
#
    print("\nWith the chosen R1,the results are:")
    Vo = R2 * Vin / (NewR1 + R2)
    print("R1 = %3.2F R2 = %3.2f Vin = %3.2f Vo = %3.3f" %(NewR1,R2,Vin,Vo))
#
# Check if happy with the values ?
#
    happy = input("\nAre you happy with the values? ")
    happy = happy.lower()
    if happy == 'y':
      break
    else:
      mode = input("Do you want to try again? ")
      mode = mode.lower()
      if mode == 'y':
        R1flag = 1
      else:
        R1flag = 0
        break
```

Figure 3.16: Program listing.

Figure 3.17 shows a typical run of the program.

```
pi@raspberrypi:~ $ python divider.py
RESISTIVE POTENTIAL DIVIDER
===========================

Input voltage (Volts): 12
Desired output voltage (Volts): 5
Enter R2 (in Ohms): 1000

R1 = 1400.00 Ohms R2 = 1000.00 Ohms

Enter chosen R1 (Ohms): 1200

With the chosen R1,the results are:
R1 = 1200.00 R2 = 1000.00 Vin = 12.00 Vo = 5.455

Are you happy with the values? y
pi@raspberrypi:~ $
```

Figure 3.17: Typical run of the program.

3.8 Resistive Attenuator Design

A resistive attenuator is a two-port resistive circuit designed to attenuate the voltage and hence the power applied to its input terminals. Attenuators are generally used in radio, audio, and communication circuits to weaken a (too) strong signal. Attenuators are also used to provide impedance matching between a source and a load. The performance of an attenuator is expressed by the number of decibels (dB) the input signal has decreased. Expressed mathematically, the logarithm to base 10 of the ratio of the output signal to the input signal, multiplied by 20 is known as the decibel:

$$dB = 20 \log_{10} V_o / V_{in}$$

For example, if the ratio of output/input voltage is 0.707, this corresponds to –3 dB. Similarly, a ratio of 0.5 corresponds to –6 dB. To find the actual voltage ratio with a given dB, we have to take the anti-logarithm as given by the following formula:

$$V_o / V_{in} = \text{antilog}_{10} (dB / 20)$$

As an example, –10dB corresponds to the voltage ratio of $V_o / V_{in} = \text{antilog}_{10} (-0.5) = 0.316$.

In this example, you get to calculate the resistances in a resistive attenuator circuit where the source and the load resistances may or may not be equal to each other. In this section, you will only consider a 'pi' (п) type network shown in Figure 3.18.

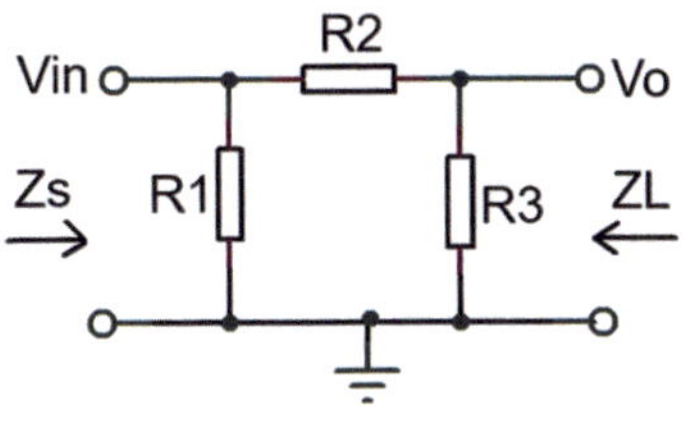

Figure 3.18: Pi type attenuator.

The design equations are as shown in Figure 3.19.

$$R_1 = Z_S\left(\frac{K^2 - 1}{K^2 - 2K\sqrt{\frac{Z_S}{Z_L}} + 1}\right)$$

$$R_2 = 0.5\sqrt{Z_S \times Z_L}\left(\frac{K^2 - 1}{K}\right)$$

$$R_3 = Z_L\left(\frac{K^2 - 1}{K^2 - \frac{2K}{\sqrt{Z_S \div Z_L}} + 1}\right)$$

Figure 3.19: Pi attenuator design equations.

Program listing: Figure 3.20 shows the program listing (program: **attenuator.py**). After displaying a heading, the user is prompted to enter the source and the load resistances Z_S and Z_L, and the attenuation factor **K** in decibels. The program calculates and displays the values of the resistors.

```
#-----------------------------------------------------
#              RESISTIVE Pi ATTENUATOR DESIGN
#              ===============================
#
# This program designs a Pi type resistive attenuator
# where the source and the load resistances may be
# different to each other
#
# Author: Dogan Ibrahim
# File  : attenuator.py
# Date  : February, 2024
#-----------------------------------------------------
import math

print("Resistive Attenuator Design")
print("===========================")
ZS = float(input("Enter source resistance in Ohms: "))
ZL = float(input("Enter load resistance in Ohms: "))
Kdb = float(input("Enter attenuation Factor in dB: "))
K = pow(10.0, Kdb/20.0)
```

```
R1 = ZS * ((K * K - 1) / (K * K - 2 * K * math.sqrt(ZS / ZL) + 1))
R2 = 0.5 * (math.sqrt(ZS * ZL)) * (K * K - 1) / K
R3 = ZL * ((K * K - 1) / ( 1 + K * K - (2 * K / math.sqrt(ZS/ZL))))

print("\nResults:")
print("-------")
print("ZS = %f Ohms\nZL = %f Ohms\nK = %f dB\nK (Vi/Vo) = %f\nR1=%f Ohms\nR2=%f
Ohms\nR3=%f Ohms"%(ZS, ZL, Kdb, K, R1, R2, R3))
```

Figure 3.20: Program listing.

An example run of the program is shown in Figure 3.21. The design parameters are as follows:

source resistance, Z_S = 75 ohms
load resistance, Z_L = 50 ohms
attenuation factor = 6 decibels

```
pi@raspberrypi:~ $ python attenuator.py
Resistive Attenuator Design
============================
Enter source resistance in Ohms: 75
Enter load resistance in Ohms: 50
Enter attenuation Factor in dB: 6

Results:
-------
ZS = 75.000000 Ohms
ZL = 50.000000 Ohms
K = 6.000000 dB
K (Vi/Vo) = 1.995262
R1=2386.203034 Ohms
R2=45.746520 Ohms
R3=86.517113 Ohms
pi@raspberrypi:~ $
```

Figure 3.21: Example run of the program.

The designed circuit is shown in Figure 3.22.

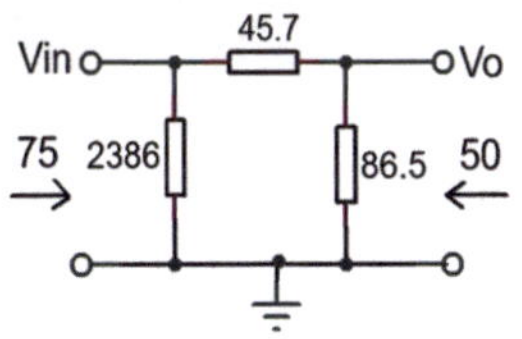

Figure 3.22: Designed attenuator circuit.

3.9 RC Charging Transient Circuit Response

RC circuits are used in many radio and communications circuits. A typical RC transient circuit consists of a resistor in series with a capacitor as shown in Figure 3.23. When the switch is closed, the voltage across the capacitor rises exponentially with a time constant, T = RC.

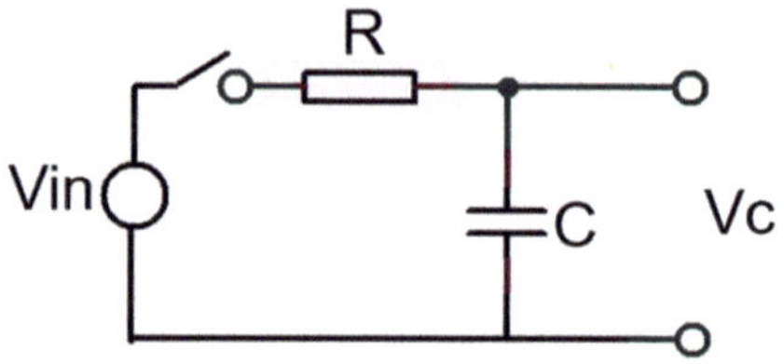

Figure 3.23: Charging RC circuit.

Expressed mathematically, assuming that initially the capacitor is discharged, when the switch is closed the voltage across the capacitor rises a given by the following formula:

$$Vc = Vin\left(1 - e^{-t/RC}\right)$$

Initially, the voltage across the capacitor is 0 V, and in steady state the voltage across the capacitor becomes equal to V_{in}. The time constant is the time where the output voltage rises to around 63.2% of its final value.

Program Listing: This example uses the Matplotlib Python plotting library that can be used to create two dimensional graphs. Before using this package, it has to be installed on your Raspberry Pi 5 using the following command:

pi@raspberrypi:~ $ **sudo apt-get install python3-matplotlib**

You must import module matplotlib at the beginning of you programs using the statement:

import matplotlib.pyplot as plt

Figure 3.24 shows the program listing (program: **RCrise.py**). After displaying heading, the values of the input voltage V_{in}, and resistor and capacitor values are read from the keyboard. The program then calculates the time constant as **T=RC** and displays the time constant and also draws the time response of the circuit. The graph is drawn as the time value (x-axis) changes from 0 to 6T and 50 points are taken to draw the graph. The time constant is also written on the graph at the point (Time constant, V_{in} / 2). The horizontal axis is in seconds, while the vertical axis is in volts.

```
#-----------------------------------------------
#              RC TRANSIENT RESPONSE
#              =====================
#
# This program reads the R and C values and then
# calculates and displays the time conctant. Also,
# the time response of teh circuit is drawn
#
# Author: Dogan Ibrahim
# File  : RCrise.py
```

```
# Date  : February, 2024
#-----------------------------------------------
import matplotlib.pyplot as plt
import numpy as np
import math

print("RC Transient Response")
print("=====================")

#
# Read Vin, R and C
#
Vin = float(input("Enter Vin in Volts: "))
R = float(input("Enter R in Ohms: "))
C = float(input("Enter C in microfarads: "))
C = C / 1000000.0

#
# Calculate and display time constant
#
T = R * C
F = 6.0 * T
N = F / 50.0
print("Time constant = %f seconds" %(T))

#
# Now plot the time response
#
x = np.arange(0, F, N)
y = [(Vin * (1.0 - math.exp(-i/T))) for i in x]

plt.plot(x, y)
plt.xlabel("Time (s)")
plt.ylabel("Capacitor Volts")
plt.title("RC Response")
plt.grid(True)
TC = "T="+str(T)+"s"
plt.text(T, Vin/2, TC)
plt.show()
```

Figure 3.24: Program listing.

Figure 3.25 shows an example graph displayed by the program. Notice that the program must run in Desktop GUI mode (e.g., using the **Terminal** menu). In this program the following input values were used (see Figure 3.26):

V_{in} = 10 volts
R = 100 ohms
C = 10 microfarads

The time constant was calculated to be 0.1 seconds.

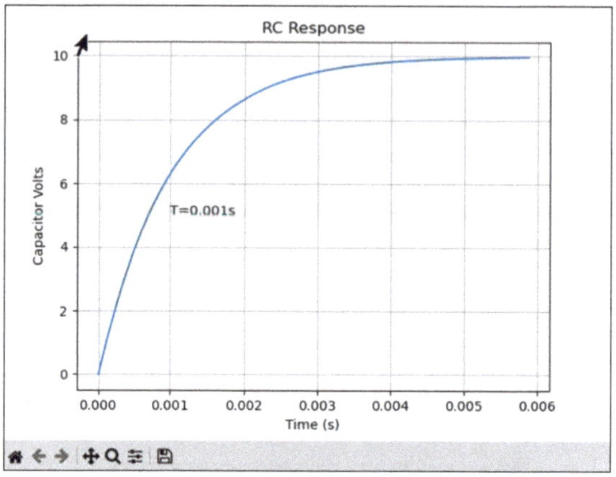

Figure 3.25: Graph plotted by the program.

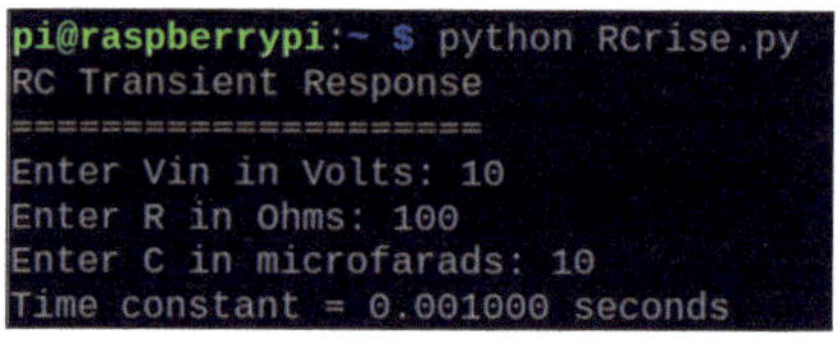

Figure 3.26: Input values to the example program.

3.10 Calculating the Inductance of a Single-Layer, Air-Core Coil

Single layer coils are used in almost all communications equipment. In this example, we will calculate the inductance of an air core single layer coil, given its diameter, length, and the number of turns.

The inductance of a single-layer air-core coil can be calculated using the well-known Wheeler's formula. The formula given below is accurate to within 1% for $L/D > 0.4$ i.e., given that the coil is not too short. For short coils, this formula is not very suitable:

$$L = \frac{D^2N^2}{45D+100Len}$$

Where *L* is the inductance in microhenries, *N* is the number of turns, *D* is the diameter of the coil in cm, and *Len* is the length of the coil in cm.

Program listing: Figure 3.27 shows the program listing (program: **coil.py**). The program reads D, N, and L from the keyboard. The inductance is then calculated and displayed. The user is warned that the results may not be accurate if the coil is short. Figure 3.28 shows a typical run of the program.

```
#==========================================================================
#                Inductance of a Single Layer Coil
#                =================================
#
# This program calculates the inductance of a single layer air
# core coil. Number of turns,diameter,and length are read from keyboard
#
# Author: Dogan Ibrahim
# File  : coil.py
# Date  : February, 2024
#==========================================================================

N = float(input("Enter Number of turns: "))
D = float(input("Enter Diameter (cm): "))
L = float(input("Enter Length (cm): "))


#
# Now carry out the calculation and display the inductance
#
Ind = (D * D * N * N) / (45 * D + 100* L)
print("Inductance = %10.4f" %(Ind))
if L / D <= 0.4:
  print("Coil is too short, may not be accurate")
else:
  print("Inductance (microhenries)= %10.4f" %(Ind))
```

Figure 3.27: Program listing.

```
pi@raspberrypi:~ $ python coil.py
Enter Number of turns: 100
Enter Diameter (cm): 2
Enter Length (cm): 15
Inductance =     25.1572
Inductance (microhenries)=     25.1572
pi@raspberrypi:~ $ █
```

Figure 3.28: Typical run of the program.

3.11 Constructing a single layer coil for required inductance

In this example, we will read the coil diameter *D*, coil length *Len*, and the required inductance and then calculate the required number of turns *N*. Also, the diameter of the required wire is displayed, assuming the coil is wound closely with the wires touching each other.

The coil design equation can be written as:

$$N^2 = \frac{(45D+100Len)L}{D}$$

The required wire diameter is given by *Len/N*.

Program listing: Figure 3.29 shows the program listing (program: **coil2.py**). The program reads *D*, *Len*, and *L* and calculates and displays *N* and the wire diameter. An example run of the program is given in Figure 3.30.

```
#========================================================================
#               Inductance of a Single Layer Coil
#               =================================
#
# This program calculates the required number of turns of a single
# layer air core coil. Also, the wire diameter is calculated
#
# Author: Dogan Ibrahim
# File  : coil2.py
# Date  : February, 2024
#========================================================================
import math

L = float(input("Enter inductance (microhenries): "))
D = float(input("Enter Diameter (cm): "))
Len = float(input("Enter Length (cm): "))

#
# Now carry out the calculation and display N
#
N = (45 * D + 100 * Len) * L / (D * D)
N = math.sqrt(N)
print("Number of turns = %10.4f" %(N))
W = 10 * Len / N
print("Wire diameter (mm) = %10.4f" %(W))
```

Figure 3.29: Program listing.

```
Enter inductance (microhenries): 50
Enter Diameter (cm): 3
Enter Length (cm): 5
Number of turns =    59.3951
Wire diameter (mm) =     0.8418
```

Figure 3.30: Example run of the program.

3.12 Calculating the Capacitance for Required Resonance Frequency

In a series resonance, RLC circuit there is a frequency point where the capacitive reactance of the capacitor becomes equal to the inductive reactance of the inductor. The point at which this occurs is called the **resonant frequency** of the circuit. Series-resonance circuits are used in tuned circuits in communication systems, mains filters, noise filters etc. The resonant frequency f_r of an RLC circuit is given by the following formula:

$$f_r = \frac{1}{2\pi\sqrt{LC}}$$

Given the required frequency and inductor values, the required capacitor value can be calculated from the following formula:

$$C = \frac{1}{4\pi^2 f^2 L}$$

Program listing: Figure 3.31 shows the program listing (program: **rescap.py**). The required frequency (in Hz) and inductor value (in mH) are entered. The program calculates the required capacitor value in μF.

```
#===============================================================
#              Capacitor value in Resonance
#              ============================
#
# This program reads the required frequency and inductor values
# and calculates the capacitor value
#
# Author: Dogan Ibrahim
# File  : rescap.py
# Date  : February, 2024
#===============================================================
import math
pi = 3.14159

#
# Read the circuit components
#
L = float(input("Enter L (in mH): "))
f = float(input("Enter f (in Hz): "))

L = L /1000.0                          # convert to Henries


#
# Now carry out the calculations
#
C = 1 / (4 * pi * pi * L * f * f)
Cuf = C * 1000000.0

print("C(in uF) = %10.2f" %(Cuf))
```

Figure 3.31: Program listing.

Figure 3.32 shows an example run of the program.

```
pi@raspberrypi:~ $ python rescap.py
Enter L (in mH): 100
Enter f (in Hz): 1000
C(in uF) =       0.25
pi@raspberrypi:~ $
```

Figure 3.32: Example run of the program.

3.13 DC Circuits Mesh Analysis

Mesh analysis is a method of representing an electrical circuit as a matrix and then analyzing and calculating the current and voltages in the circuit. In this example, we will analyze a simple DC circuit consisting of three resistors and two batteries. It is assumed that the readers are familiar with the basic circuit theory and have analyzed circuits using the mesh analysis before.

In mesh analysis, we assume currents in each loop of the circuit and then enter the circuit elements (resistors and batteries in this case). Then the matrix equation *Ax = b* is formed where *A* represent the circuit elements in terms of resistances, *x* represents the currents in the circuit, and *b* represents the voltage sources (e.g., batteries). The matrix equation is then solved as shown in the examples earlier in this chapter. After calculating the currents in the circuit, we can easily calculate the voltage across every resistor.

In this example, the circuit shown in Figure 3.33 is used as an example. This circuit consists of 3 resistors: 10 ohms, 20 ohms, and 40 ohms, and two voltage sources (batteries): 10 volts and 20 volts.

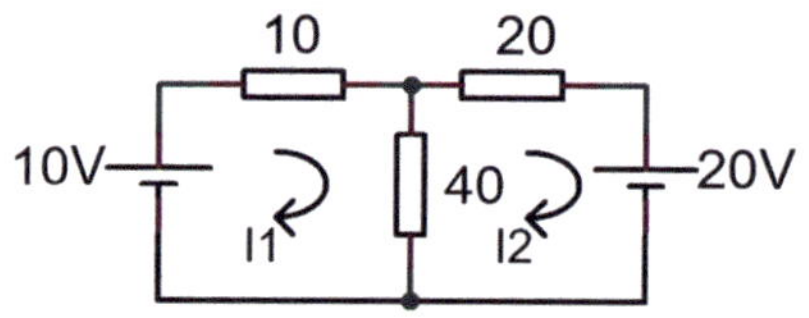

Figure 3.33: Circuit to be analyzed.

From an inspection of the circuit, we can write the elements of matrices *A*, *b*, and *x* for this circuit are as follows:

Here,

$$A = \begin{bmatrix} 50 & -40 \\ -40 & 60 \end{bmatrix}$$

$$x = \begin{bmatrix} I1 \\ I2 \end{bmatrix}$$

$$b = \begin{bmatrix} 10 \\ -20 \end{bmatrix}$$

Just as a reminder, assuming that the currents in each loop are taken in the same direction (e.g., clockwise):

A_{11} = Sum of all the resistances in loop1.
A_{12} = Sum of all the resistances common to loop 1 and loop 2 (negative).
A_{21} = Sum of all the resistances common to loop 2 and loop 1 (negative).
A_{22} = Sum of all the resistances in loop 2.

Program Listing: Figure 3.34 shows the program listing (program: **mesh.py**). At the beginning of the program, a heading is displayed. Arrays *A* and *b* are cleared (filled with zeroes). The program then reads the number of rows of the *A* matrix, which is also equal to the number of columns. Then the elements of *A* and *b* matrices are read and the currents are calculated and displayed. As shown in Figure 3.35, the currents in the two loops are: I_1 = -0.143 A and I_2 = -0.429 A. We can then calculate the voltage across each resistor as follows:

Voltage across 10 ohm resistor = $10 \times I_1 = -1.43$ volts
Voltage across 40 ohm resistor = $40 \times (I_1 - I_2) = 11.44$ volts
Voltage across 20 ohm resistor = $20 \times I_2 = -8.58$ volts

```
#=====================================================
#              CIRCUIT MESH ANALYSIS
#              =====================
#
# This program carries out mesh analysis of a circuit.
# The user enters the circuit elements in the form of
# A and b matrices and the program calculates and
# displays the values of the currents in the circuit
#
# Author: Dogan Ibrahim
# File  : mesh.py
# Date  : February, 2023
#=======================================================
from numpy import array
from numpy import *
from numpy.linalg import solve

print("Circuit Mesh Analysis")
print("=====================")
rows = int(input("Enter number of rows of A: "))
cols = rows
A = zeros((rows, cols))
b = zeros(rows)

#
# Read the matrix A elements
```

```
#
for i in range(0, rows):
  for j in range(0, cols):
    A[i, j] = float(input("Enter A%d%d:" %(i+1, j+1)))

#
# Read the matrix b elements
#
print("")
for i in range(0, rows):
  b[i] = float(input("Enter b%d:" %(i+1)))

#
# Calculate and display the currents in the circuit
#
x = solve(A, b)

print("\nResults:")
for i in range(0, rows):
  print("I%d = %7.3fA" %(i+1, x[i]))
```

Figure 3.34: Program listing.

```
pi@raspberrypi:~ $ python mesh.py
Circuit Mesh Analysis
======================
Enter number of rows of A: 2
Enter A11:50
Enter A12:-40
Enter A21:-40
Enter A22:60

Enter b1:10
Enter b2:-20

Results:
I1 =  -0.143A
I2 =  -0.429A
```

Figure 3.35: Calculating the currents in the circuit.

3.13.1 DC Circuits mesh analysis — a more complex example

In this example, a more complex circuit is given which includes a current source in addition to voltage sources. The currents in the circuit are calculated using the program **mesh.py**.

The circuit used in this example is shown in Figure 3.36 — it consists of four resistors, a voltage source, and a current source.

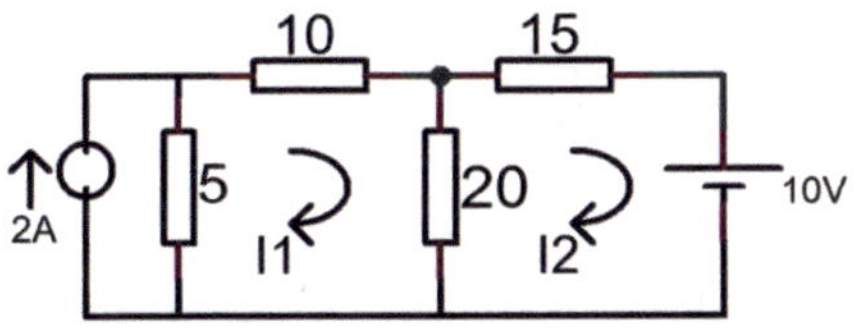

Figure 3.36: Circuit for the example.

The circuit shown in Figure 3.36 can be converted into its equivalent circuit by replacing the current source with voltage source as shown in Figure 3.37.

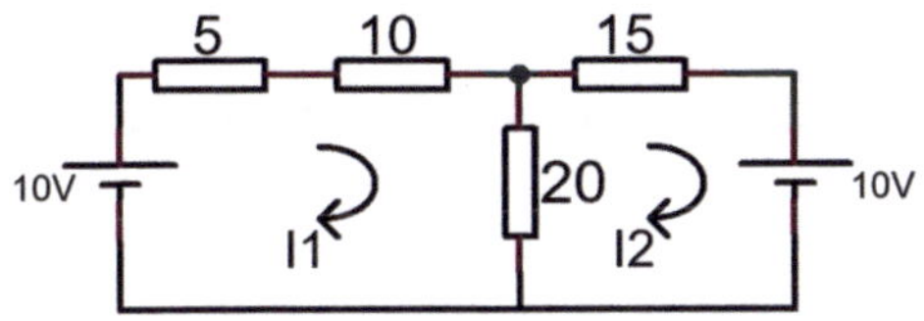

Figure 3.37: Equivalent circuit of Figure 3.36.

The circuit now consists of two loops with 4 resistors. Using the mesh analysis program, the current in the circuit are calculated to be: I_1 = 0.182 A, I_2 = –0.182 A as shown in Figure 3.38.

```
pi@raspberrypi:~ $ python mesh.py
Circuit Mesh Analysis
=====================
Enter number of rows of A: 2
Enter A11:35
Enter A12:-20
Enter A21:-20
Enter A22:35

Enter b1:10
Enter b2:-10

Results:
I1 =   0.182A
I2 =  -0.182A
```

Figure 3.38: Calculating currents in the circuit.

3.14 DC Circuit Node Analysis

In this example, we analyze a simple circuit using the node analysis method. Node analysis gives the voltages at the node points of the circuit.

Node analysis is similar to mesh analysis but here you have to enter the circuit elements at the nodes of the circuit. If there are *n* nodes in a circuit then there will be *n-1* independent equations where one node is taken to be the reference node. Then the matrix equation **Ax = b** is formed where *A* represent the circuit elements in terms of conductance (inverse or resistance), *x* represents the voltages at each node in the circuit, and *b* represents the current sources. The matrix equation is then solved as shown in the examples earlier in this chapter. After calculating the voltages in the circuit, we can easily calculate the current through every resistor.

In this section, the circuit shown in Figure 3.39 is used as an example. This circuit consists of four resistors and two current sources, and has three nodes.

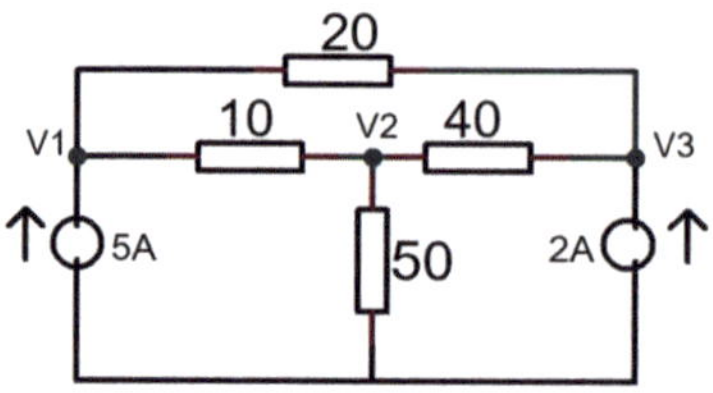

Figure 3.39: Circuit to be analyzed.

From an inspection of the circuit, you can write the elements of matrices *A*, *b*, and *x* for this circuit are as follows:

Here,

$$A = \begin{vmatrix} \frac{1}{10}+1/20 & -1/10 & -1/20 \\ -1/10 & \frac{1}{10}+\frac{1}{40}+1/50 & -1/40 \\ -1/20 & -1/40 & \frac{1}{20}+1/40 \end{vmatrix}$$

$$x = \begin{bmatrix} V1 \\ V2 \\ V3 \end{bmatrix}$$

$$b = \begin{bmatrix} 5 \\ 0 \\ 2 \end{bmatrix}$$

Note. To make the calculation simpler, we will be entering the resistances to the program and the program will convert them into conductance for its internal calculations.

Figure 3.40 shows the program listing (program: **node.py**). At the beginning of the program a heading is displayed and the user is asked to enter the number of nodes there are in the circuit (excluding the reference node). The elements at each node are then entered for the *A* matrix in terms of the resistances at each node. **Note** that a '0 'must be entered to terminate the entry for a node. The program internally converts the entered values into conductance by taking their inverses. This makes it easier to enter the element values. The currents at each node are then entered for the *b* matrix. The program calculates and displays the voltages at the nodes. After calculating the voltages, we can easily calculate the current through each resistor by dividing the voltage across it by its resistance.

```
#=====================================================
#              CIRCUIT NODE ANALYSIS
#              =====================
#
# This program carries out node analysis of a circuit.
# The user enters the circuit elements in the form of
# A and b matrices and the program calculates and
# displays the values of the voltages in the circuit
#
# Author: Dogan Ibrahim
# File  : node.py
# Date  : February, 2024
#=======================================================
from numpy import array
from numpy import *
from numpy.linalg import solve

print("Circuit Node Analysis")
print("=====================")
rows = int(input("Enter number of nodes: "))
cols = rows
A = zeros((rows, cols))
b = zeros(rows)

#
# Read the matrix A elements
#
total = 0
elements = 1
for i in range(0, rows):
  for j in range(0, cols):
    while True:
       elements = float(input("Enter A%d%d:" %(i+1, j+1)))
       if elements == 0:
         break
       total = total + (1 / elements)
       A[i, j] = total
    total = 0
    print("")

#
# Read the matrix b elements
#
print("")
for i in range(0, rows):
  b[i] = float(input("Enter b%d:" %(i+1)))
```

```
#
# Calculate and display the currents in the circuit
#
x = solve(A, b)

print("\nResults:")
for i in range(0, rows):
  print("V%d = %7.3fV" %(i+1, x[i]))
```

Figure 3.40: Program listing.

Figure 3.41 shows the output when the program is run for this example. The voltages at each node are calculated to be: V_1 = 404.286 V, V_2 = 350.000 V, V_3 = 412.875 V.

```
Circuit Node Analysis
=====================
Enter number of nodes: 3
Enter A11:10
Enter A11:20
Enter A11:0

Enter A12:-10
Enter A12:0

Enter A13:-20
Enter A13:0

Enter A21:-10
Enter A21:0

Enter A22:10
Enter A22:40
Enter A22:50
Enter A22:0

Enter A23:-40
Enter A23:0

Enter A31:-20
Enter A31:0

Enter A32:-40
Enter A32:0

Enter A33:20
Enter A33:40
Enter A33:0


Enter b1:5
Enter b2:0
Enter b3:2

Results:
V1 = 404.286A
V2 = 350.000A
V3 = 412.857A
```

Figure 3.41: Calculating voltages in the circuit.

3.15 Bipolar Junction Transistor Analysis

In this example, we will analyze the DC voltages and currents in a BJT transistor circuit shown in Figure 3.42. The user enters the values of all the resistors, supply voltage, and the value of the transistor current gain β.

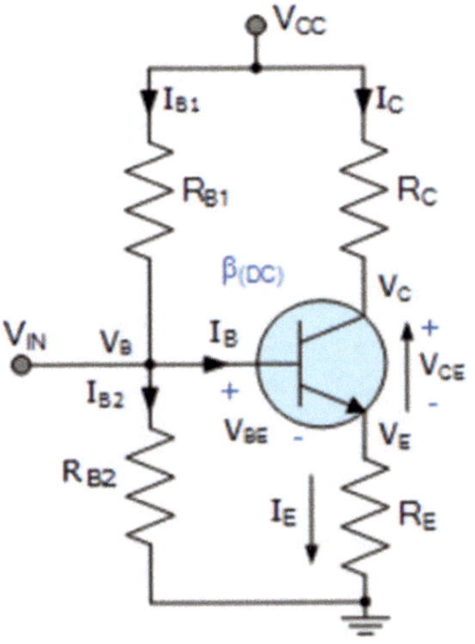

Figure 3.42: BJT transistor circuit.

The voltages and currents relating to Figure 3.42 can be calculated as follows:

$$V_B = V_{cc}\left(\frac{R_{B2}}{R_{B1}+R_{B2}}\right)$$

$$I_{B1} = \frac{V_{cc}-V_B}{R_{B1}}$$

$$I_E = \frac{(V_B-0.7)}{R_E}$$

$$R_B = \frac{R_{B1}R_{B2}}{R_{B1}+R_{B2}}$$

$$I_B = \frac{V_B-0.7}{R_B+(\beta+1)R_E}$$

$$I_{B2} = I_{B1}-I_B$$

$$I_c = I_E - I_B$$

$$V_{CE} = V_{CC} - (I_c R_c + I_E R_E)$$

$$V_E = I_E R_E$$

$$I_c = \beta I_B$$

Notice that the base current is usually very small compared to the collector current and in most calculations equations one can make the approximation:

$$I_c \cong I_E$$

Figure 3.43 shows the program listing (program: **bias.py**). The program displays a heading and then reads the values of all the two resistors, supply voltage, and the current gain β (or B) of the transistor. It then calculates and displays all the voltages and current in the circuit. A typical run of the program is shown in Figure 3.44 with the following component values:

V_{CC} = 12 volts
R_{B1} = 30,000 ohms
R_{B2} = 10,000 ohms
R_C = 1000 ohms
R_E = 1000 ohms
β = 100

```
#====================================================
#              Voltage Divider Biasing
#              =======================
#
# This program analyzes a voltage divider bias
#
# Author: Dogan Ibrahim
# File  : bias.py
# Date  : February, 2024
#====================================================
print("Voltage Divider Biassed Transistor Analysis")
print("===========================================")
#
# Read the component values
#
Vcc = float(input("Enter supply voltage (Volts): "))
RB1 = float(input("Enter RB1 (Ohms): "))
RB2 = float(input("Enter RB2 (ohms): "))
RC = float(input("Enter RC (Ohms): "))
RE = float(input("Enter RE (Ohms): "))
B = float(input("Enter current gain(B): "))

#
# Calculations
#
Vb = 0.7
VB = Vcc * (RB2 / (RB1 + RB2))
IB1 = (Vcc - VB) / RB1
IE = (VB - Vb) / RE
RB = RB1 * RB2 / (RB1 + RB2)
IB = (VB - Vb) / (RB + RE * (B + 1))
IB2 = IB1 - IB
IC = IE - IB
VCE =  Vcc - (IC * RC + IE * RE)
```

```
VE = IE * RE
VC = Vcc - RC * IC
IB1 = 1000.0 * IB1
IB2 = 1000.0 * IB2
IE = 1000.0 * IE
IC = 1000.0 * IC
IB = 1000.0 * IB

#
# Display the results
#
print("\nVB=%7.3f V\nVCE=%7.3f V\nVE=%7.3f V\nVC=%7.3f V" %(VB,VCE,VE,VC))
print("\nIB1=%7.3f mA\nIB2=%7.3f mA\nIB=%7.3f mA\nIC=%7.3f mA\nIE=%7.3f mA"
%(IB1,IB2,IB,IC,IE))
```

Figure 3.43: Program listing.

```
pi@raspberrypi:~ $ python bias.py
Voltage Divider Biassed Transistor Analysis
===========================================
Enter supply voltage (Volts): 12
Enter RB1 (Ohms): 30000
Enter RB2 (ohms): 10000
Enter RC (Ohms): 1000
Enter RE (Ohms): 1000
Enter current gain(B): 100

VB=  3.000 V
VCE=  7.421 V
VE=  2.300 V
VC=  9.721 V

IB1=  0.300 mA
IB2=  0.279 mA
IB=  0.021 mA
IC=  2.279 mA
IE=  2.300 mA
pi@raspberrypi:~ $ 
```

Figure 3.44 Typical run of the program.

3.16 Designing Active Low-Pass Filters

Low-pass filters are commonly used in almost all communications circuits. In this example, we will design a second order low-pass active filter. Users enter the cut-off frequency of the filter and the program will calculate and display the component values.

There are several forms of second order active low-pass filters. The one used in this example study is the well-known Sallen-Key type filter whose circuit diagram is shown in Figure 3.45. The filter consists of two capacitors and two resistors. This type of filter has unity gain (i.e., 1) in the pass-band.

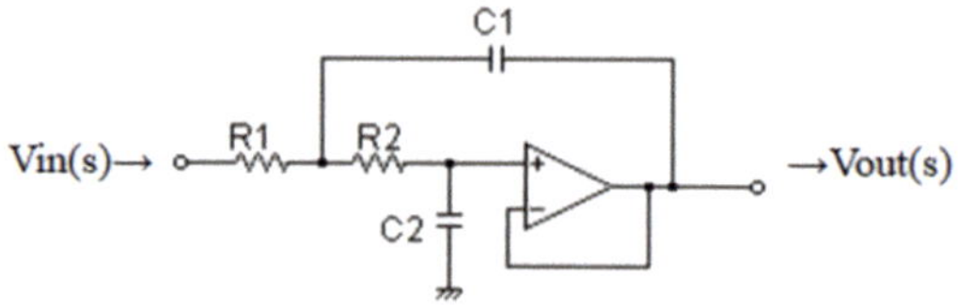

Figure 3.45: Second-order low-pass filter.

Although it is possible to design Sallen-Key type filters with higher gains, care should be taken since the high-gain filters can easily become unstable and oscillate, giving high overshoot in their time responses and high gains at their cut-off frequencies. The *Q* factor of a filter defines its quality, i.e., the damping of the response. For low-pass filters, *Q* should have a value of 0.707. Higher values of *Q* may cause instability and overshoots in the time and frequency responses. Another parameter used in filter design is the damping factor, denoted with ξ. This is related to the *Q* factor with the following equation:

$$\xi = \frac{1}{2Q}$$

The ideal value for ξ is also 0.707. Lower values of ξ give rise to oscillations and instability in the designed circuit. To maintain stability the gain of an active filter must not be greater than 3. The relationship between the gain and *Q* factor is:

$$G = 3 - \frac{1}{Q}$$

Figure 3.46 shows the filter frequency response for various values of ξ. It is clear from this figure that ξ = 0.707 gives a flat response.

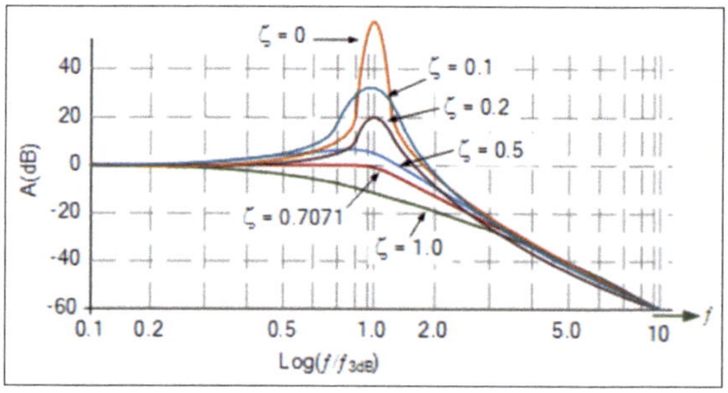

Figure 3.46: Filter frequency response for different values of ξ.

The cut-off frequency of the filter in Figure 3.45 is given by:

$$f = \frac{1}{2\pi\sqrt{R_1 R_2 C_1 C_2}}$$

In order to make sure that $Q = 0.707$, the following must be satisfied:

Let $R_1 = R_2 = R$, and $C_1 / C_2 = n$, then we have to make sure that $C_1 / C_2 >= 4Q^2$, i.e., $C_1 / C_2 >= 2$,

Or $C_2 / C_1 <= 0.5$.

The filter design equation then becomes:

$$f = \frac{1}{2\pi R C_2 \sqrt{n}}$$

or

$$R = \frac{1}{2\pi f C_2 \sqrt{n}}$$

The design steps are then as follows:

- Read the filter cut-off frequency.
- Select a value for $C2$.
- Select a value for $C1$, making sure that $C_1 / C_2 >= 2$. For $Q = 0.707$, select $C_1 / C_2 = 2$.
- Calculate the required R.

Figure 3.47 shows the program listing (program: **lowpass.py**). At the beginning of the program a heading is displayed and the user is required to enter the filter cut-off frequency, C_1 and C_2 in microfarads. The program calculates and displays the required resistors, making sure that the above inequality is satisfied. The user is then asked to enter the actual physical resistors to be used and displays the actual cut-off frequency when these resistors are used. If the user is not happy with the calculated cut-off frequency, then the program is re-started so that the user can select new values for the capacitors.

```
#=======================================================
#    Second Order Low-Pass Active Filter Design
#    ==========================================
#
# This program designs a second order low-pass filter
#
# Author: Dogan Ibrahim
# File  : lowpass.py
# Date  : February, 2024
#=======================================================
import math

yn = 'n'
while yn == 'n':
  print("Second Order Low-Pass Active Filter Design")
  print("==========================================")
#
```

```
# Read the cut-off frequency and C1, C2
#
  C1 = 1
  C2 = 1
  f = float(input("Enter the cut-off frequency (Hz): "))

  while C1/C2 < 2:
    C2 = float(input("Enter C2 (microfards): "))
    C = 2 * C2
    C1 = float(input("Enter C1 (microfarads) C1 >= %d : " %(C)))

  C1F = C1 / 1000000.0
  C2F = C2 / 1000000.0
#
# Calculations
#
  n = C1F / C2F
  r = 2 * math.pi * f * C2F * math.sqrt(n)
  R = 1 / r

#
# Display the results
#
  print("\nRESULTS:")
  print("=======")
  print("C1=%7.3f uF\nC2=%7.3f uF\nR1=%7.3f Ohm\nR2=%7.3f Ohm" %(C1,C2,R,R))

  print("\nEnter the actual physical resistors to be used:")
  Rnew = float(input("Enter R1, R2 (Ohms): "))
  fn = 2 * math.pi * Rnew * C2F * math.sqrt(n)
  fnew = 1 / fn
  print("Actual cut-off requency = %7.3f Hz" %(fnew))
  yn = input("Are you happy with the actual cut-off frequecy (yn)? ")
  yn = yn.lower()
```

Figure 3.47: Program listing.

An example run of the program is shown in Figure 3.48. In this example, the following specifications were used:

Cut-off frequency = 1000 Hz
$C2$ = 12 µF
$C1$ = 6 µF

```
Second Order Low-Pass Active Filter Design
==========================================
Enter the cut-off frequency (Hz): 1000
Enter C2 (microfards): 6
Enter C1 (microfarads) C1 >= 12 : 12

RESULTS:
=======
C1= 12.000 uF
C2=  6.000 uF
R1= 18.757 Ohm
R2= 18.757 Ohm

Enter the actual physical resistors to be used:
Enter R1, R2 (Ohms): 18
Actual cut-off requency = 1042.033 Hz
Are you happy with the actual cut-off frequecy (yn)? y
>>> |
```

Figure 3.48: Example run of the program.

The resistor values are calculated to be $R_1 = R_2 = 18.759$ ohms. Selecting the nearest physical value as 18 ohms gives the cut-off frequency as 1042.033 Hz. We accept this value and respond with **y** to terminate the program. The application given in the following link can be used to display the frequency response, *Q* factor etc of a low-pass Sallen-Key type filter:

http://sim.okawa-denshi.jp/en/OPstool.php

Using this application, the transfer function, *Q*, the damping factor, the frequency and phase responses of the designed filter with the new resistors is shown in Figure 3.49 and Figure 3.50.

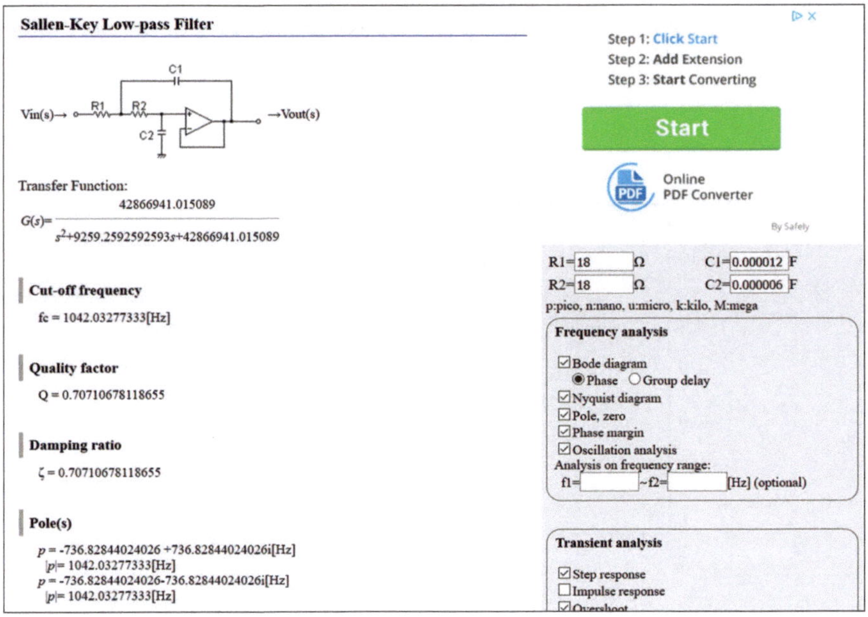

Figure 3.49: Designed filter specifications.

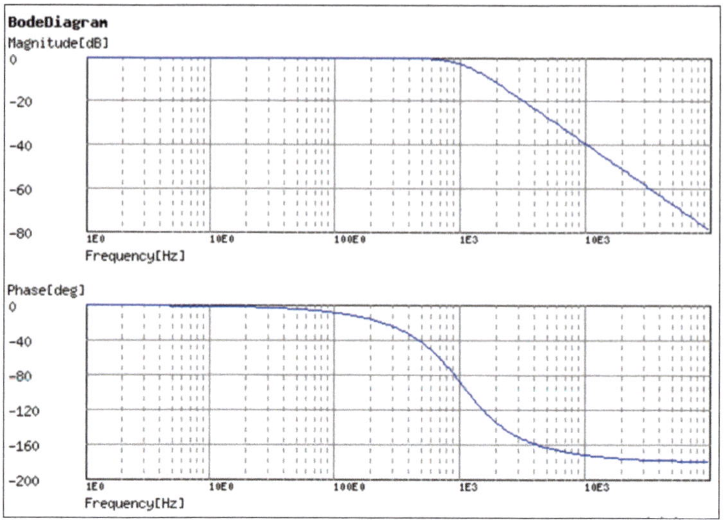

Figure 3.50: Frequency and phase responses of the designed filter.

You can design higher-order filters by cascading the basic second order section given in this example.

3.17 Passive Low-Pass Butterworth Filter Design

Passive filters are used in many communications circuits. In this section, a program is presented to design low-pass Butterworth type passive filters having capacitors and inductors. Figure 3.51 shows the frequency response of low-pass Butterworth filters with different orders.

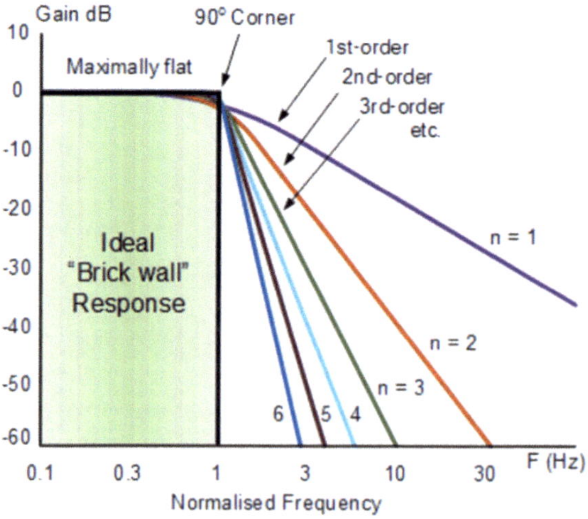

Figure 3.51: Frequency response of Butterworth filters

Low-pass Butterworth filters can be realized using the **Cauer topology**, where inductors and capacitors are used in the topology. The topology assumes 1 ohm source and load resistances and 1 rad/sec frequency. As shown below, there are two topologies depending whether an inductor or a shunt capacitor is first used at the source end.

Type 1

Figure 3.52 shows the Type-1 topology.

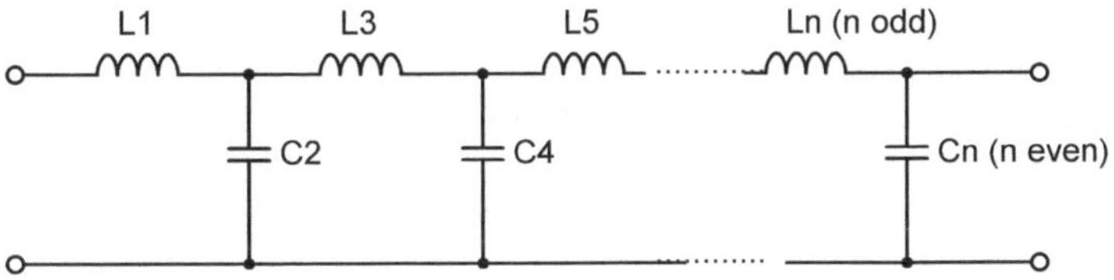

Figure 3.52: Filter Type-1 topology.

The coefficients of the Butterworth low-pass filter are given by:

$$C_k = 2\sin\left[\frac{(2k-1)}{2n}\pi\right]$$ where k is even.

$$L_k = 2\sin\left[\frac{(2k-1)}{2n}\pi\right]$$ where k is odd.

Type 2

Figure 3.53 shows the Type-2 topology.

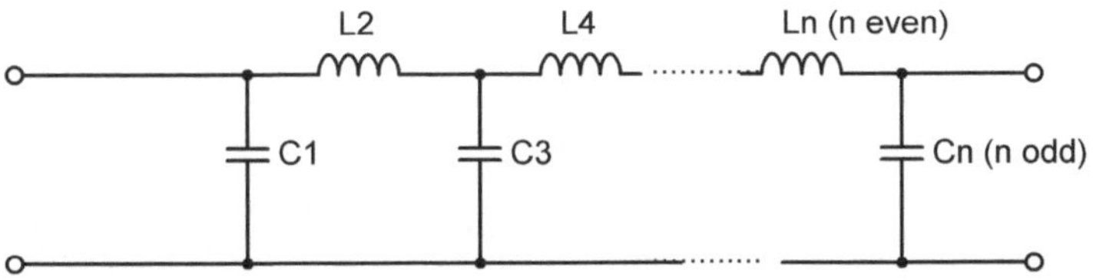

Figure 3.53: Filter Type-2 topology.

The coefficients of the Butterworth low-pass filter are given by:

$$C_k = 2\sin\left[\frac{(2k-1)}{2n}\pi\right]$$ where k is odd.

$$L_k = 2\sin\left[\frac{(2k-1)}{2n}\pi\right]$$ where k is even.

After finding the *L* and *C* values, we have to scale them to the required source resistance, load resistance, and the cut-off frequency by applying the following transformations:

$L \rightarrow L \times R / \omega_c$
$C \rightarrow C / (\omega_c \times R)$

R is the source and load resistance (they are equal), and $\omega_c = 2 \times \pi \times f$ where *f* is the required cut-off frequency.

Figure 3.54 shows the program listing (program: **LPPassive**). At the beginning of the program, the user is required to select the topology type and then enter the required source and load resistance, order of the filter, and then the required cut-off frequency. The program calculates and displays the L and C values for the chosen topology.

```
#=====================================================================
#              PASSIVE BUTTERWORTH LOW-PASS FILTER DESIGN
#
# This program displays the L and C values for a low-pass filter
# whose order, cut-off frequency and source (load) resistance given
#
# Author: Dogan Ibrahim
# File  : LPPassive.py
# Date  : February, 2024
#=====================================================================
import math
print("Passive Butterworth Low-Pass Filter Design")
topology = int(input("Choose topology Type 1 (1) or Type 2 (2): "))
n = int(input("Enter filter order: "))
R = int(input("Enter source and load resistance (Ohms): "))
f = int(input("Enter cut-off frequency (Hz): "))
w = 2 * 3.14159 * f

if topology == 1:
   for k in range(1, n+1):
      m = (2 * k - 1) * 3.14159 / ( 2 * n)
      if (k % 2) == 0:
         C = 2 * math.sin(m)
         CC = 1000000 * C / (w * R)
         print("C%d = %f uF" %(k, CC))
      else:
         L = 2 * math.sin(m)
         LL = 1000 * L * R / w
         print("L%d = %f mH" %(k, LL))
elif topology == 2:
   for k in range(1, n+1):
      m = (2 * k - 1) * 3.14159 / ( 2 * n)
```

```
    if (k % 2) == 0:
        L = 2 * math.sin(m)
        LL = 1000 * L * R / w
        print("L%d = %f mH" %(k, LL))
    else:
        C = 2 * math.sin(m)
        CC = 1000000 * C / (w * R)
        print("C%d = %f uF" %(k, CC))
```

Figure 3.54: Program listing.

Figure 3.55 shows an example output of designing a Type 1 4th order Butterworth passive low-pass filter with a cut-off frequency of 1000 Hz and source (load) resistance of 50 ohms. The designed filter circuit is shown in Figure 3.56.

```
Passive Butterworth Low-Pass Filter Design
Choose topology Type 1 (1) or Type 2 (2): 1
Enter filter order: 4
Enter source and load resistance (Ohms): 50
Enter cut-off frequency (Hz): 1000
L1 = 6.090597 mH
C2 = 5.881603 uF
L3 = 14.704022 mH
C4 = 2.436254 uF

>>> |
```

Figure 3.55: Example output.

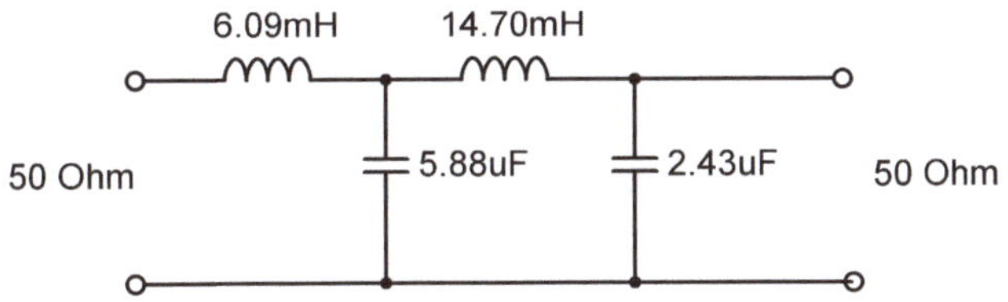

Figure 3.56: Designed filter circuit.

3.18 The 555 Timer IC

LM555/NE555 is a very popular integrated circuit used in monostable and astable circuit designs. The chip is commonly used as a monostable to generate delays, or as an astable to generate square wave signals. In this section, you will give a program that can be used to design 555-based monostable and astable circuits. Note that LM556/NE556 is a dual timer chip, including two 555 timer modules.

Monostable circuit

Figure 3.57 shows the circuit diagram of a 555 based monostable circuit. A negative-going (high-to-low) trigger pulse is applied to pin 2 of the chip. A pulse is generated at output pin 3. The duration of this pulse is set by R and C by the following formula:

$$T = 1.1RC$$

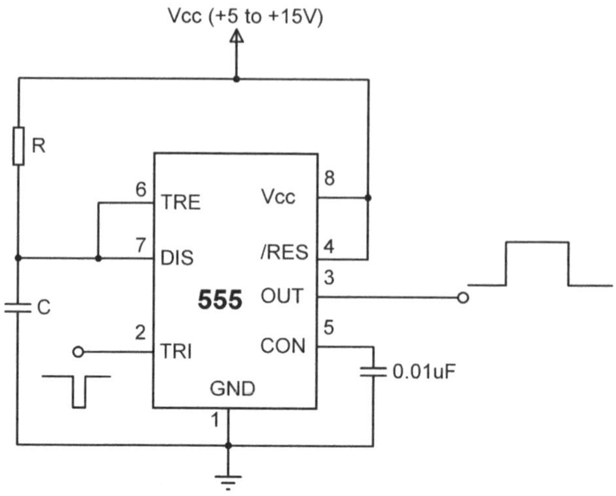

Figure 3.57: Monostable circuit.

Astable circuit

Figure 3.58 shows the circuit diagram of a 555 based astable circuit. The period T of the output square waveform is set by R1, R2 and C1 and is given by:

$T = T_h + T_l$, where T_h is the output high time and T_l is output low time.

and

$T_h = 0.7 \times (R1 + R2) \times C1$
$T_l = 0.7 \times R2 \times C1$

Using the above formula, the period of the output waveform is given by:

$T = 0.7 \times (R1 + 2R2) \times C1$

And the frequency is given by:

$$f = \frac{1}{T} = \frac{1.44}{(R1 + 2R2)C1}$$

The **duty cycle** of the waveform is the ratio of the ON time to the period and is given by:

$D = 100 \times T_h / T$ %

or $D = (R1 + R2) / (R1 + 2R2) \times 100\%$

Note that 50% duty cycle is not possible. For a duty cycle as close as feasible to 50%, choose a large value for R2.

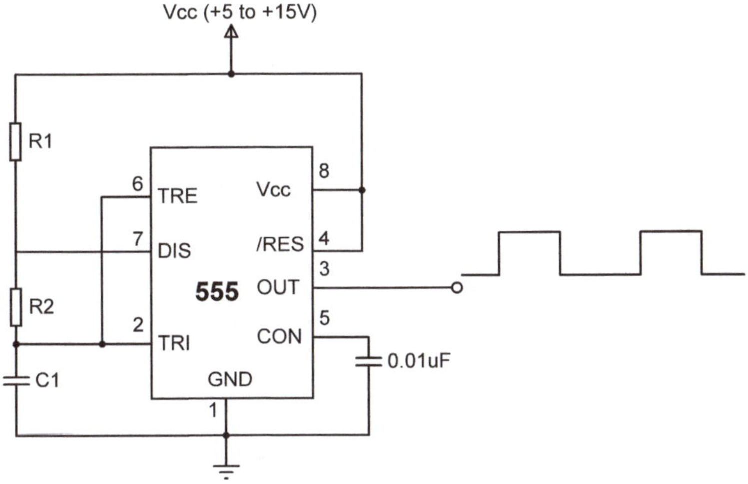

Figure 3.58: Astable circuit.

Figure 3.59 shows the program listing (program: **NE555**). At the beginning of the program user is asked to enter m (monostable) or a (astable). If a monostable circuit is to be designed, then **t** and **C** values are read and **R** is calculated and displayed. If an astable circuit is to be designed, then **f**, **C1**, and **R2** values are read and **R1** is calculated and displayed.

```
#==========================================================================
#              NE555/LM555 TIMER CHIP
#
# This program calculate steh component values for designing monostable
# and astable circuits using a 555 chip
#
# Author: Dogan Ibrahim
# File: NE555.py
# Date: February, 2024
#===========================================================================
print("555 MONOSTABLE-ASTABLE DESIGN")
mode = input("Monostable(m) or Astable (a) ?: ")
if mode == 'm':
    t = float(input("Required output pulse width (seconds): "))
    C = float(input("Enter capacitor C (uF): "))
    C = C / 1000000
    R = t / (1.1 * C)
    print("Required resistor value is (Ohms): %5.2f" %(R))
elif mode == 'a':
    f = float(input("Required frequency (Hz): "))
    C1 = float(input("Enter capacitor C1 (uF): "))
    C1 = C1 / 1000000
    R2 = float(input("Enter R2 (Ohms): "))
```

```
R1 = 1.44 / (f * C1) - 2 * R2
print("Required resistor R1 value is (Ohms): %5.2f" %(R1))
D = 100 * (R1 + R2) / (R1 + 2 * R2)
print("Duty cycle = %3.2f percent" %(D))
```

Figure 3.59: Program listing.

Figure 3.60 shows design of an astable circuit with frequency 1 Hz, C = 1 μF, and R2 chosen as 680 kilo-ohms.

```
pi@raspberrypi:~ $ python NE555.py
555 MONOSTABLE-ASTABLE DESIGN
Monostable(m) or Astable (a) ?: a
Required frequency (Hz): 1
Enter capacitor C1 (uF): 1
Enter R2 (Ohms): 680000
Required resistor R1 value is (Ohms): 80000.00
Duty cycle = 52.78 percent
pi@raspberrypi:~ $
```

Figure 3.60: Example design of an astable circuit.

3.19 Impedance Matching

Impedance matching is very important concept in maximum power transfer applications. For maximum power transfer the source and the load resistances must be the same. In RF design we usually have complex source and/or load impedances with real parts and reactances. Such impedances are usually matched at a given frequency. Matching complex impedances is not an easy task. In this section, a program is given which can be used to calculate the required inductor and capacitor values in order to match a source impedance to a load impedance at a given frequency.

In this section, L-type matching networks are considered. Figure 3.61 (**Shunt-series**) and Figure 3.62 (**Series-shunt**) show two types of impedance matching circuits. Here, (R1 + *j*X1)is the source impedance, and (R2 + *j*X2) is the load impedance.

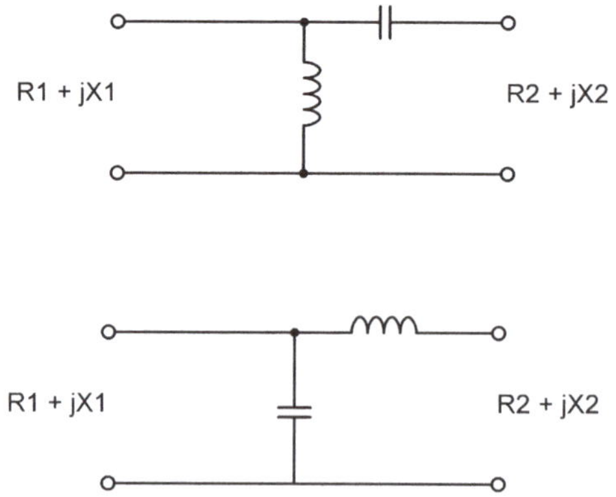

Figure 3.61: Shunt-series impedance matching circuit.

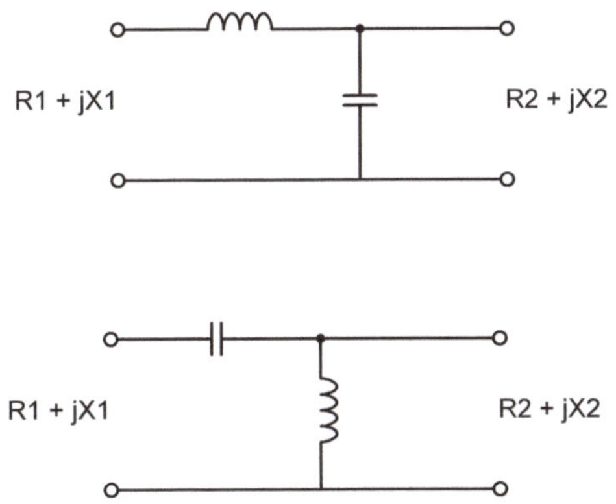

Figure 3.62: Series-shunt impedance matching circuit.

Impedance matching is a complex process and interested readers can get the required formula from the following site (Francesco Urbani):

https://urbanij.github.io/projects/matching_networks/

Figure 3.63 shows the program listing (program: **match**). At the beginning of the program the user enters the source and load impedances and the frequency of operation. The program calculates the matching circuit components and displays them on the screen.

```
#=====================================================================
#               IMPEDANCE MATCHING
#
# This program calculates the inductance and capacitance for
# impedance matching a source to a load at a given frequency
#
# Author: Dogan Ibrahim
# Date  : February, 2024
# File  : match.py
#=====================================================================
R1=float(input("R1 (Ohm): "))
X1=float(input("X1 (Ohm): "))
R2=float(input("R2 (Ohm): "))
X2=float(input("X2 (Ohm): "))
f=float(input("freq (MHz): "))
w = 2 * 3141592 * f

if R1 * (R1 - R2) + X1 * X1 >= 0:
   print("Shunt-Series Design")
   xshu1=(R1*X2+R2*X1-R1*(X2-((R2*(R1**2-R2*R1+X1**2))/R1)**(1/2)))/(R1-R2)
   xser1=X2-((R2*(R1**2-R2*R1+X1**2))/R1)**(1/2)
   xshu2=(R1*X2+R2*X1-R1*(X2+((R2*(R1**2-R2*R1+X1**2))/R1)**(1/2)))/(R1-R2)
   xser2=X2+((R2*(R1**2-R2*R1+X1**2))/R1)**(1/2)
```

```
    Lshunt = 1E9 * xshu1 / w
    print("shunt inductor=%f Ohm, %5.3f nH" %(xshu1, Lshunt))

    Cseries = 1E12 / (w * xser2)
    print("series capacitor=%f Ohm, %5.3f pF" %(xser2, Cseries))

    print("OR")
    Cshunt = -1E12 / (w * xshu2)
    print("shunt capacitor=%f Ohm, %5.3f pF" %(xshu2, Cshunt))

    Lseries = -1E9 * xser1 / w
    print("series inductor=%f Ohm, %5.3f nH" %(xser1, Lseries))

elif R2 * (R2 - R1) + X2 * X2 >= 0:
    print("Series-shunt Design")
    xshu1=(R1*X2+(R1*R2*(R2**2-R1*R2+X2**2))**(1/2))/(R1-R2)
    xser1=-(R2*X1-(R1*R2*(R2**2-R1*R2+X2**2))**(1/2))/R2
    xshu2=(R1*X2-(R1*R2*(R2**2-R1*R2+X2**2))**(1/2))/(R1-R2)
    xser2=-(R2*X1+(R1*R2*(R2**2-R1*R2+X2**2))**(1/2))/R2

    Lshunt = -1E9 * xshu1 / w
    print("shunt inductor=%f Ohm, %5.3f nH" %(xshu1, Lshunt))

    Cseries = -1E12 / (w * xser2)
    print("series capacitor=%f Ohm, %5.3f pF" %(xser2, Cseries))

    print("OR")
    Cshunt = 1E12 / (w * xshu2)
    print("shunt capacitor=%f Ohm, %5.3f pF" %(xshu2, Cshunt))

    Lseries = 1E9 * xser1 / w
    print("series inductor=%f Ohm, %5.3f nH" %(xser1, Lseries))
```

Figure 3.63: Program listing.

Figure 3.64 shows an example design. Notice that there are two solutions to the problem and both solutions are displayed by the program. The designed circuits are shown in Figure 3.65.

```
R1 (Ohm): 100
X1 (Ohm): 12
R2 (Ohm): 50
X2 (Ohm): 25
freq (MHz): 10
Shunt-Series Design
shunt inductor=113.429785 Ohm, 1805.292 nH
series capacitor=75.714889 Ohm, 210.203 pF
OR
shunt capacitor=-89.429779 Ohm, 177.966 pF
series inductor=-25.714891 Ohm, 409.265 nH

>>> |
```

Figure 3.64: Example design.

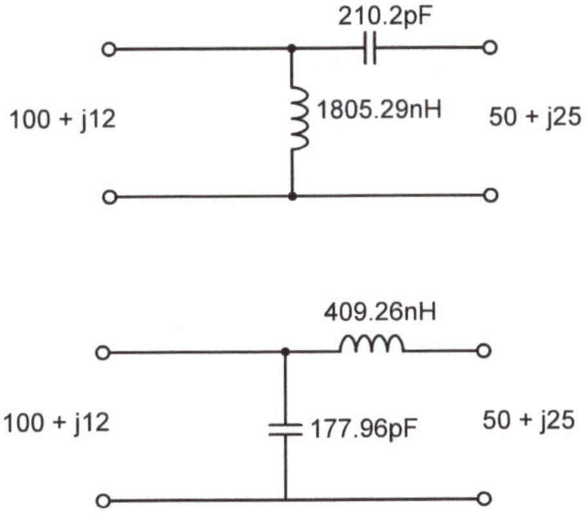

Figure 3.65: Designed impedance matching circuits.

3.20 Designing a Common-Emitter BJT Transistor Amplifier Circuit

In this example you will design a single common-emitter transistor amplifier circuit with the given supply voltage and required voltage gain. The circuit diagram of a single stage common-emitter amplifier circuit is shown in Figure 3.66. The AC parameters of this amplifier circuit can be calculated using the following formula (see Hybrid Pi model):

$$G = \frac{V_{out}}{Vin} = G_1 \times G_2 = \frac{V_{out}}{V_b} \times \frac{V_b}{V_{¿}}$$

$$G_1 = \frac{V_{out}}{V_b} = \frac{-R_c \vee ¿ R_L}{r_e}$$

$$\mathfrak{R} = \frac{25}{I_E}$$

$$R_B = \frac{R_1 R_2}{R_1 + R_2}$$

$$C_E = \frac{1}{2\pi f X_c} \text{ where } X_c = \frac{R_E}{10} \text{ at the required -3db frequency}$$

$$Z_i = \frac{R_B \beta r_e}{R_B + \beta r_e}$$

$$G_2 = \frac{V_b}{V_¿} = \frac{Z_i}{R_S + Z_i}$$

$$Z_o = R_c \vee ¿ R_L$$

$$C_2 = \frac{1}{2\pi f X_c} \text{ where, } X_c = \frac{Z_o}{10} \text{ at the required -3db frequency}$$

Notice that in general, assuming that the only load resistance is R_L and $I_E \approx I_C$, the voltage gain is given by:

$A_v = -g_m R_L$

Where, $g_m = I_c / V_T$

and $V_T = kT / q$

At room temperature (15 degrees C), V_T = 25mV.

Therefore,

$A_v = -R_L \times I_c / 25$

The design starts from the specification of the required voltage gain, or the input voltage and the required output voltage. The power supply voltage is usually supplied. We first have to find the resistor values to for biasing the transistor and also providing the required gain. Briefly, the steps are as follows:

- Given the input and required output voltages, calculate the required voltage gain.
- Choose a collector current of around 2 mA.
- Assume the V_{CE} to be $V_{CC}/2$.
- Choose a value for R_L. If the gain without the source and load resistors is given by R_L x I_E /25 at room temperature, choose R_L to give 30% higher gain than the required value.
- Choose a value for R_E (assuming that $I_E = I_c$) from: $Vcc = I_cR_L + V_{CE} + I_cR_E$.
- Calculate base current from I_B / β.
- Calculate the base voltage from $V_B = 0.7 + I_cR_E$
- Assume that the current in R_1 and R_2 will be 10 times the base current and calculate suitable values for R_1 and R_2.
- Calculate the input and output impedances.

- Check that the overall gain is as required, if not go back and adjust I_c, R_L, or R_E.
- Calculate suitable values for C_1, C_2 and C_E.

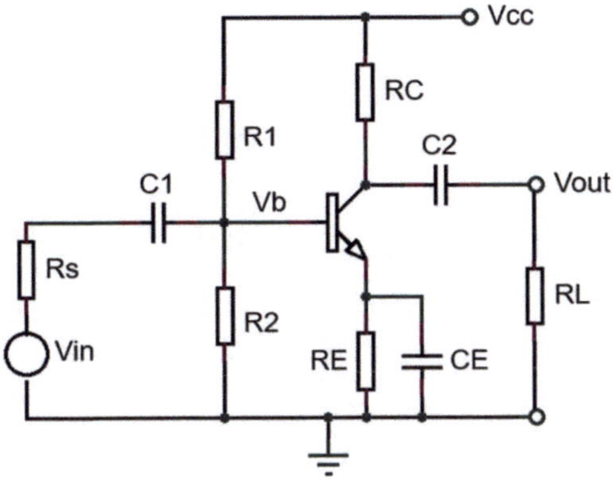

Figure 3.66: Common-emitter amplifier circuit.

Figure 3.67 shows the program listing (program: **amplifier.py**). At the beginning of the program, a heading is displayed. The program then reads the various parameters and requirements. The program calculates the parameters listed above and displays on the monitor.

```
#=======================================================
#    Single Stage Common Emitter Amplifier Design
#    ==============================================
#
# This program designs a common emitter amplifier
#
#
# Author: Dogan Ibrahim
# File  : amplifier.py
# Date  : February, 2024
#========================================================
import math

print("Single Stage Common Emitter Amplifier Design")
print("============================================")
#
# Read the component values
#
Vcc = float(input("Enter supply voltage (Volts): "))
Vin = float(input("Enter Vin (Volts): "))
Vout = float(input("Enter Vout (Volts): "))
RS = float(input("Enter Rs (ohms): "))
RL = float(input("Enter RL (Ohms): "))
```

```
f = float(input("Enter lowest frequency to amplify (Hz): "))
B = float(input("Enter current ratio (B): "))

#
# Calculations
#
Gr = Vout / Vin                         # Gain
IC = 2                                  # IC=2mA
Ge = Gr * (1.3)                         # Estimate gain 30% higher
RC = Ge * 25 / IC                       # Collector resistance
RCRL = (RC * RL) / (RC + RL)
re = 25.0 / IC
Rinbase = B * re                        # Base resistance
G1 = RCRL / re                          # G1
VCE = Vcc / 2                           # VCE
IC = IC / 1000.0
RE = (Vcc - RC * IC - VCE) / IC         # RE
VB = 0.7 + IC * RE
IB = IC / B
IR1R2 = 10.0 * IB                       # 10 times base current
R2 = VB / IR1R2
R1 = (Vcc - VB) / IR1R2
RB = (R1 * R2) / (R1 + R2)
Zi = (RB * Rinbase) / (RB + Rinbase)
G2 = Zi / (RS + Zi)
G = G1 * G2
XC = Zi / 10.0
C1 = 1000000.0 / (2 * math.pi * f * XC)
Zo = RCRL
XC = Zo / 10.0
C2 = 1000000.0 / (2 * math.pi * f * XC)
XC = RE / 10.0
CE = 1000000.0 / (2 * math.pi * f * XC)

#
# Display the results
#
print("\nRESULTS:")
print("=======")
print("Gain=%7.3f\nIC=%7.3f mA\nIB=%7.3f mA" %(G,IC*1000.0,IB*1000.0))
print("\nVcc=%7.3f V\nVB=%7.3f V\nVCE=%7.3f V" %(Vcc,VB,VCE))
print("\nR1=%7.3f Ohm\nR2=%7.3f Ohm\nRC=%7.3f Ohm\nRE=%7.3f Ohm" %(R1,R2,RC,RE))
print("\nC1=%7.3f uF\nC2=%7.3f uF\nCE=%7.3f uF" %(C1,C2,CE))
print("\nZi=%7.3f Ohm\nZo=%7.3f Ohm" %(Zi,Zo))
```

Figure 3.67: Program listing.

An example run of the program is shown in Figure 3.68. In this example, the following specifications were assumed:

V_{in} = 0.010 V
V_{out} = 0.4 V
R_s = 100 ohms
R_L = 15 kilo-ohms
f_l = 20 kHz
V_{cc} = 12 V
β = 100

It is clear that the required voltage gain of the transistor (without the source and load resistances) is 0.4 V / 0.010 V = 40. We can use the following physical components for our circuit:

R1 = 33 kΩ
R2 = 27 kΩ
RC = 680 Ω
RE = 2.2 kΩ
C1 = C2 = C3 = 1 µF

```
pi@raspberrypi:~ $ python amplifier.py
Single Stage Common Emitter Amplifier Design
============================================
Enter supply voltage (Volts): 12
Enter Vin (Volts): 0.010
Enter Vout (Volts): 0.4
Enter Rs (ohms): 120
Enter RL (Ohms): 15000
Enter lowest frequency to amplify (Hz): 20000
Enter current ratio (B): 100

RESULTS:
=======
Gain= 45.142
IC=  2.000 mA
IB=  0.020 mA

Vcc= 12.000 V
VB=  5.400 V
VCE=  6.000 V

R1=33000.000 Ohm
R2=27000.000 Ohm
RC=650.000 Ohm
RE=2350.000 Ohm

C1=  0.069 uF
C2=  0.128 uF
CE=  0.034 uF

Zi=1152.950 Ohm
Zo=623.003 Ohm
pi@raspberrypi:~ $
```

Figure 3.68: Example run of the program.

3.21 Using a Windows-Based Simulation Program

Simulation is a powerful tool that enables an electrical circuit to be analyzed on a PC. Using a simulator program, you can draw a circuit on a PC and then display all the voltages and currents in the circuit, draw waveforms, etc. Simulation can be used for passive and active electrical circuits as well as for digital circuits.

Although the simulation is an invaluable tool in designing and developing electronic circuits, it has the following advantages and disadvantages:

- Any component whatever the cost is, can be modelled and simulated using a simulator. The virtual instruments are computer programs and thus there are no cost issues.
- Simulation does not usually take into consideration the component tolerances, aging, or temperature effects. Readers may think that all components are ideal at all times.
- Virtual instruments and components used in a simulation cannot be damaged by connecting wrongly or by applying large voltages or currents.
- There is no calibration problems associated with virtual instruments. They are available at all times and operate with the same specifications.
- Simulation allows measurements of internal currents and voltages that in many cases are virtually impossible to do while using real components.

There are many commercially available simulation programs available. There are also freely available simulation programs available on the Internet. In this book, we will be using the freely available Texas Instruments simulation package called **TINA-TI** which can be used to simulate large number of passive and active circuit as well as digital circuits.

You can use simulation to check the operation of your electronic circuits. In this section, we will use the windows based TINA-TI simulation program to compare the theoretical results of some of the circuits used in the book with the simulation results.

The steps to download the TINA-TI simulation program on your Windows PC are as follows:

- Go to the Texas Instruments web site:

  https://www.ti.com/tool/TINA-TI

- Scroll down to TINA-TI_ENGLISH and click Download (Figure 3.69)

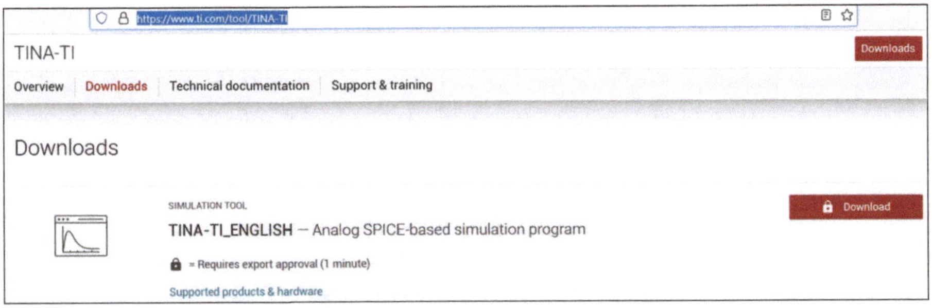

Figure 3.69: Click: Download.

- Login to the site. If this is the first time you are using the Texas Instruments site you will have to register before you can log in.
- After you logging in, click to download the TINA-TI simulation package. At the time of authoring this book, the package had the name:**TINA90-Tien.9.3.200.277.zip**.
- Copy the zip file to a new folder, extract the contents and click TINA90-Tien.exe to install the simulator program on your PC.

3.21.1 Simulating the Resistive Mesh Circuit in Figure 3.33

Let's apply TINA-TI to a the previously discussed resistive mesh setup. The steps are as follows:

- Click on the icon to start the TINA-TI simulator program.
- Click on resistor and place it on the screen. Click on the resistor again and join the three resistors as in Figure 3.33.
- Click on battery and place the 10 V and 20 V batteries. Connect the batteries to the circuit as shown in Figure 3.33.
- Right click on resistors and batteries and change their values to be as in Figure 3.33.
- Insert an earth symbol at the bottom part of the circuit. It is important that every circuit to be simulated must have an ground path.
- Figure 3.70 shows the circuit built on the simulator.

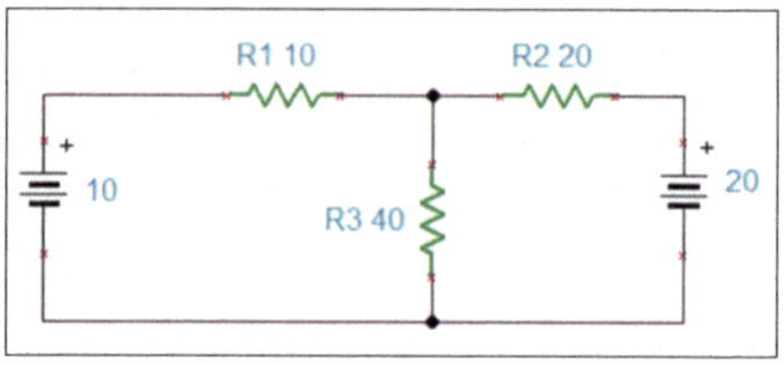

Figure 3.70 Circuit build on the simulator.

You are now ready to simulate your circuit. Click **Analysis → DC Analysis → Table of DC results**. A table of all the currents and voltages in the circuit will be displayed as shown in Figure 3.71. Note that for example, **I_R1** is the current through resistor R1. The results obtained with the simulation are the same as the theoretical results.

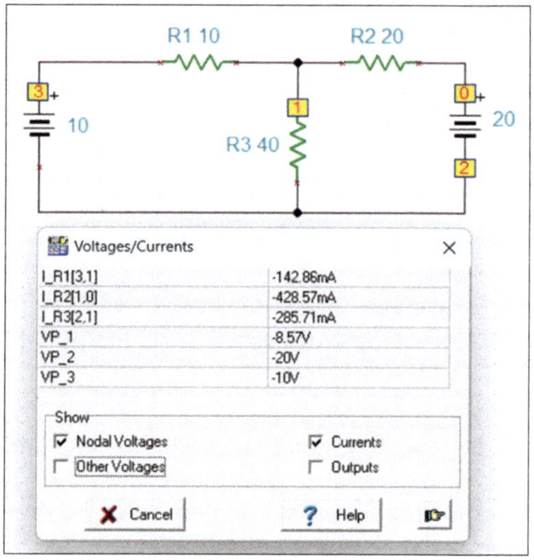

Figure 3.71: Simulation results.

3.21.2 Simulating the transistor circuit in Figure 3.42

The steps are as follows:

- Click on the icon to start the TINA-TI simulator program.
- Click on Semiconductors and place an NPN transistor.
- Place the four resistors.
- Insert a battery.
- Right click on the transistor and set the type to BC337. The minimum β of this transistor is 100.
- Right-click on the battery and set its voltage to 12 V.

- Right-click on resistors and set their values as in Figure 3.42.
- Figure 3.72 shows the circuit built on the simulator.

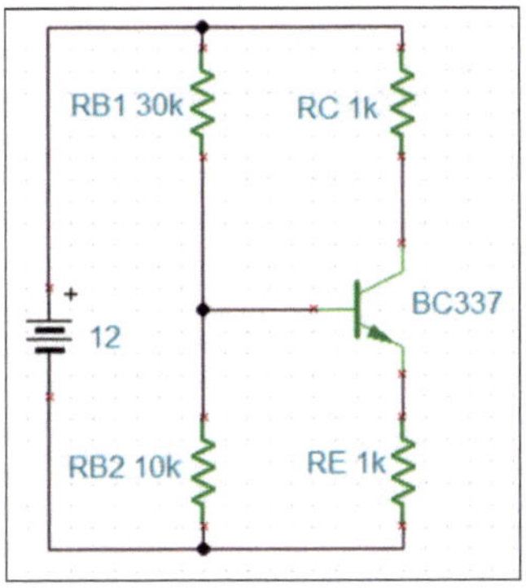

Figure 3.72: Circuit built on the simulator.

You are now ready to simulate your circuit. Click **Analysis → DC Analysis → Table of DC results**. A table of all the currents and voltages in the circuit will be displayed as shown in Figure 3.73. The results obtained with the simulation are the same as the theoretical results.

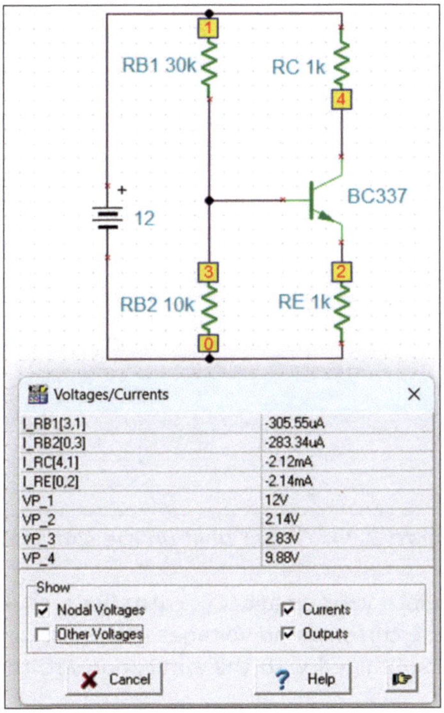

Figure 3.73: Simulation results.

3.21.3 Simulating the transistor circuit in Figure 3.66

The steps are as follows:

- Click on the icon to start the TINA-TI simulator program.
- Click on Semiconductors and place an NPN transistor.
- Place the resistors.
- Place the capacitors
- Insert a battery.
- Insert a voltage generator.
- Right click on the transistor and set the type to BC337. The minimum β of this transistor is 100.
- Right-click on the battery and set its voltage to 12 V.
- Right-click on resistors and capacitors and set their values as in Figure 3.66.
- Right click on the voltage generator and set it to sine wave with amplitude 0.010V

Figure 3.74 shows the circuit built on the simulator

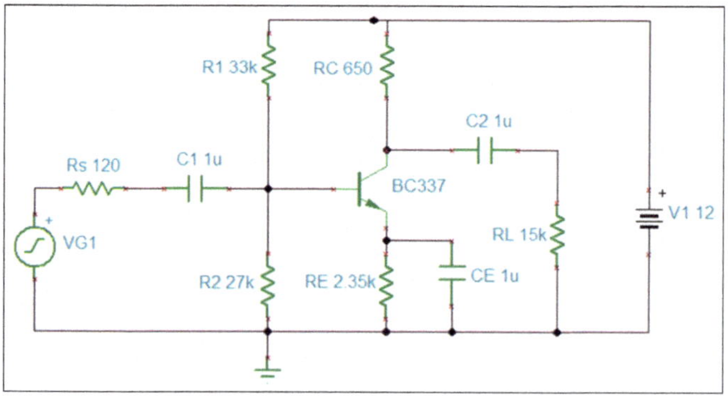

Figure 3.74: Circuit built on the simulator.

You are now ready to simulate your circuit. Click **Analysis → DC Analysis → Table of DC results**. A table of all the currents and voltages in the circuit will be displayed as shown in Figure 3.75. The results obtained with the simulation are the same as the theoretical results.

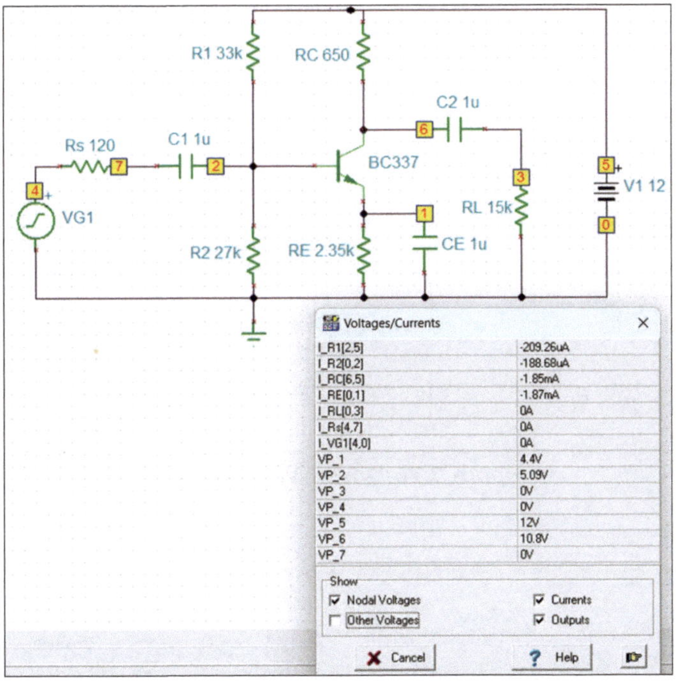

Figure 3.75: DC voltages and currents in the circuit.

Let's proceed by displaying the input and output waveforms on a virtual oscilloscope and then calculate the voltage gain of the amplifier. This requires Voltage Pins to be inserted at the input and output of the circuit. The steps are:

- Click Meters and select the Voltage Pin, then place it at the output of the circuit (named VF1).
- Click Meters again and select the Voltage Pin and place it at the input of the circuit (named VF2). See Figure 3.76.
- Click Analysis and select the Oscilloscope.
- Set the Channel to VF1 and X source to VF1, the Time/Div to 20u (20 microseconds), and Volts/Div to 500m (500 mV). You should see the amplifier output voltage displayed in red color.
- Set the Channel to VF2, the Volts/Div to 20m (20mV). You should see the amplifier input voltage displayed in green.
- The output voltage is measured to be about 800mV peak to peak. Similarly, the input voltage is measured to be about 20 mV peak to peak. Therefore, the voltage gain of the amplifier is about 800 / 20 = 40 as was calculated theoretically.

Figure 3.77 shows the input and output waveforms of the amplifier.

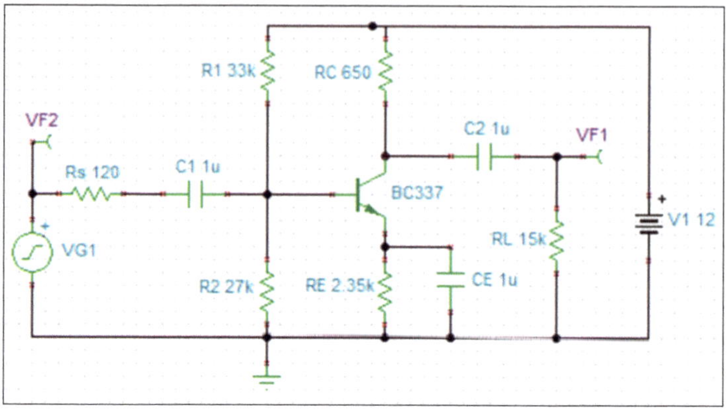

Figure 3.76: Insert Voltage Pins.

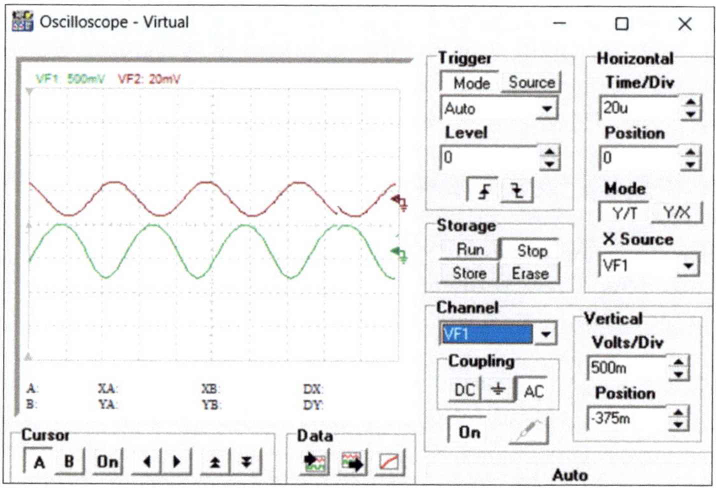

Figure 3.77: Input-output waveforms.

Chapter 4 • Hardware Based Projects for Amateur Radio

4.1 Overview

In this chapter, hardware based projects are given that could be useful to amateur radio operators. The projects are built using the Raspberry Pi 5. Note that the author Dogan Ibrahim has a book entitled *Raspberry Pi for Radio Amateurs* published by Elektor. Some of the projects in that book have been modified to run under Raspberry Pi 5 and are included here.

4.2 Project 1: Logic Probe

Description: This is a simple logic probe. A logic probe is used to indicate the logic status of an unknown digital signal. In this application, a test lead (probe) is used to detect the unknown signal and two different color LEDs are used to indicate the logic status. If for example the signal is logic 0 then the RED LED is turned ON. If on the other hand the signal is logic 1 then the GREEN LED is turned ON. The circuit also detects high-impedance state where none of the LEDs are turned ON when there is no applied logic level.

Block diagram: Figure 4.1 shows the block diagram of the project. Logic input is applied to the terminal labelled Logic Probe. Because the Raspberry Pi 5 inputs are not +5 V compatible. the logic input voltage must not be greater than +3.3 V.

Figure 4.1: Block diagram of the project.

Circuit diagram: Figure 4.2 shows the circuit diagram. In this project, a type BC108 NPN transistor is used.

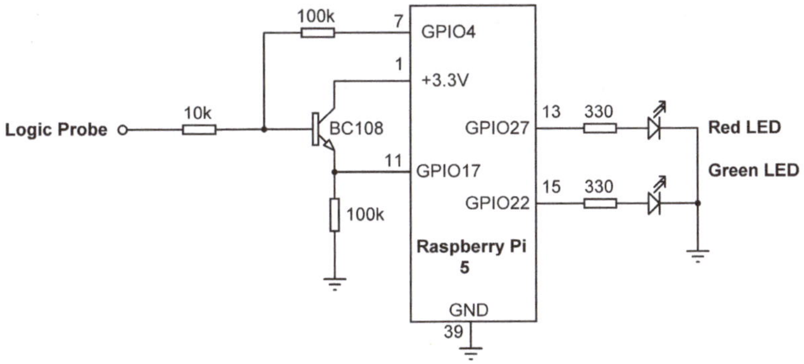

Figure 4.2: Circuit diagram of the project.

The operation of the circuit is as follows: the transistor is configured as an emitter-follower stage with the base connected to GPIO4, configured as an output port. The external logic signal is applied to the base of the transistor through a 10kΩ resistor. The emitter of the transistor is connected to GPIO17, which is configured as an input port. GPIO4 applies logic levels to the base of the transistor and then pin GPIO17 determines the state of the external signal as shown in Table 4.1. For example, if after setting **GPIO4 = 1**, we detect that **GPIO17 =1** and also after setting **GPIO4 = 0** we again detect that **GPIO17 = 1**, then the probe must be at logic 1.

Probe State	Output from GPIO4	Detected at GPIO17
At high impedance	1 0	1 0
Probe at logic 1	1 0	1 1
Probe at logic 0	1 0	0 0

Table 4.1: Applied and detected logic levels.

Program listing: Figure 4.3 shows the program listing (program: **Probe**). At the beginning of the program, **PROBE** is assigned to GPIO17, **RED** and **GREEN** assigned to GPIO27 and GPIO22 respectively, and **CHECK** assigned to GPIO4. The LEDs and CHECK are then configured as digital outputs and **PROBE** configured as digital input. The remainder of the program executes in an endless loop and implements the logic given in Table 4.1 in order to turn ON/OFF the RED and GREEN LEDs.

```
#*********************************************************************
#                       LOGIC PROBE
#                      ===========
#
# This is a logic probe program. Two LEDs, one RED and one
# GREEN are connected to Raspberry Pi 5. The program determines
# the state of the applied logic voltage. If voltage is LOW 0 then
# the RED LED is turned ON. If on the other hand the voltage is
# HIGH then the GREEN LED is turned ON.
#
# The program detects the high impedance state when there is no logic
# level applied at the input
#
#
# Autor   : Dogan Ibrahim
# Program: Probe.py
# Date    : February, 2024
#*********************************************************************
from gpiozero import LED, InputDevice, OutputDevice
from time import sleep
```

```
PROBE = InputDevice(17)
RED = LED(27)
GREEN = LED(22)
CHECK = OutputDevice(4)

#
# Turn OFF the LEDs at the start
#
RED.off()
GREEN.off()

#
# Examine the state of the applied external logic signal. If
# there is nothing connected at the input then assume high
# impedance state and turn OFF both LEDs. If on the other
# hand the state is 0, then turn ON the RED LED. if on the
# other hand the external applied logic state is 1, then
# turn ON the GREEN LED

while True:
   CHECK.on()                          # set CHECH = 1
   sleep(0.001)
   if PROBE.value == 1:
      CHECK.off()
      sleep(0.001)
      if PROBE.value == 0:
        RED.off()
        GREEN.off()                    # High impedance
      else:
        RED.off()
        GREEN.on()
   else:
      CHECK.off()
      sleep(0.001)
      if PROBE.value == 0:
        GREEN.off()
        RED.on()
```

Figure 4.3: Program listing.

The project built on a breadboard is shown in Figure 4.4.

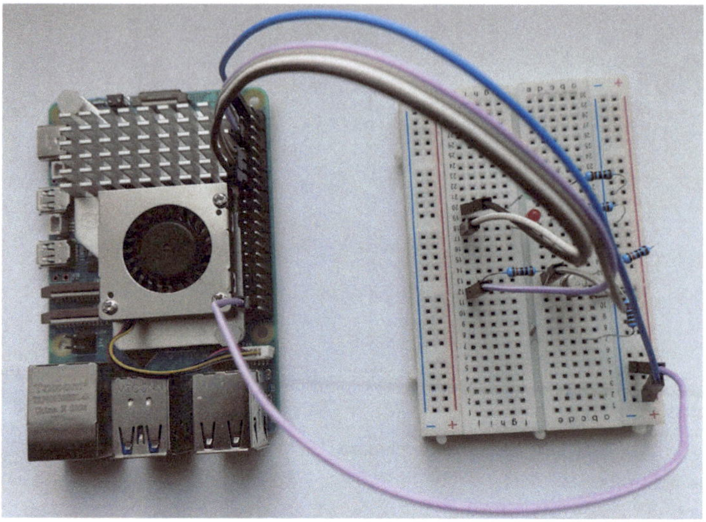

Figure 4.4: Project built on a breadboard.

4.3 Project 2: Station Mains On-Off Power Control

Description: In this project, 4 buttons and 4 relays are connected to Raspberry Pi 5. The relays change state when their corresponding buttons are pushed. For example, various mains operated equipment in the station can be connected to the mains supply through the relays, and such equipment can be switched ON or OFF easily by pressing their corresponding buttons.

Block Diagram: Figure 4.5 shows the block diagram of the project.

Figure 4.5: Block diagram of the project.

Circuit Diagram: In this project, a 4-channel relay board from Elegoo (www.elegoo.com) is used (Figure 4.6). This is an optocoupled relay board with 4 inputs — one for each channel. The relay inputs are at the bottom right hand side of the board while the relay outputs are located at the top side of the board. The middle position of each relay is the common point, the connection to its left is the normally closed (NC) contact, while the connection to the right is the normally open (NO; N.O.) contact. The relay contacts support 250VAC at 10 A and 30 VDC at 10 A. IN1, IN2, IN3 and IN4 are the **active-LOW** inputs, which means that a relay is activated when a logic LOW signal is applied to its input pin. The relay contacts are normally closed (NC). Activating the relay changes the active contacts such that the common pin and NC pin become the two relay contacts and at the same time the LED at the input circuit of the relay board corresponding to the activated relay is turned ON. The VCC can be connected to either +3.3V or to +5V. Jumper JD is used to select the voltage for the relay.

Note: because the current drawn by a relay can be in excess of 80 mA, you must remove this jumper and connect an **external power supply** (e.g., +5 V) to the JD-VCC pin.

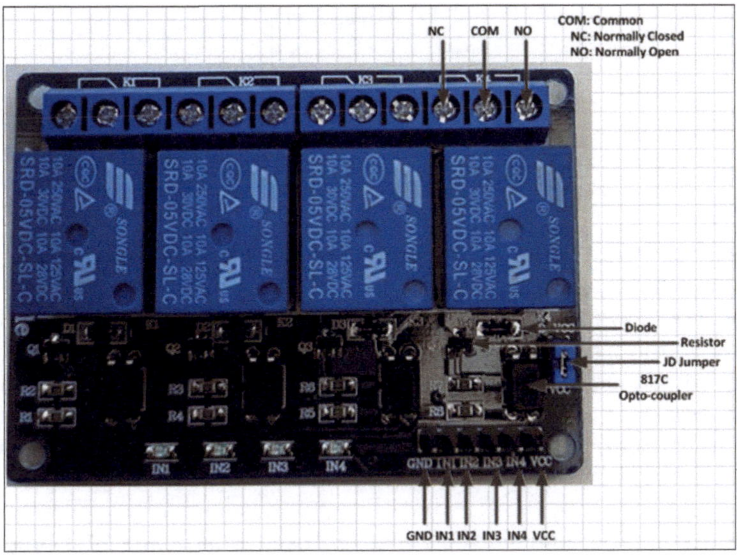

Figure 4.6: Four-channel relay board.

The circuit diagram of the project is shown in Figure 4.7. The buttons are named as A, B, C and D and they are connected to Raspberry Pi 5 port pins GPIO4, GPIO17, GPIO27, and GPIO22 respectively. The button outputs are at logic 1 and go to logic 0 when the button is pressed. Pins IN1, IN2, IN3, and IN4 of the relay module are connected to port pins GPIO21, GPIO20, GPIO16, and GPIO12 respectively. The relay contacts should be connected to the equipment to be switched ON/OFF through the mains supply.

WARNING: care should be taken when working with live mains voltages and you should seek professional advice if you are not sure about connections to the mains supply.

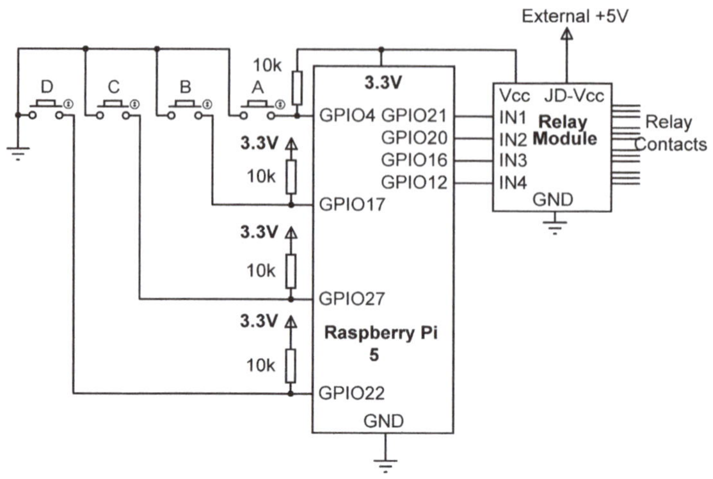

Figure 4.7: Circuit diagram of the project.

Program Listing: Figure 4.8 shows the program listing (program: **relays.py**). At the beginning of the program the button and relay connection to the Raspberry Pi are defined. Port pins where the relay input pins IN1-IN4 are connected to are configured as outputs. Port pins where the button pins BUTTONA, BUTTONB, BUTTONC, and BUTTOND are connected to are configured as inputs. All the 4 relays are then turned OFF at the beginning of the program. The remainder of the program is executed inside the main program loop where a **while** loop is used. Here, function **TestButton** is used to check the state of a button and if the button is pressed then the state of the corresponding relay is toggled.

Functions are useful programming tools that are used when parts of a program is to be repeated several times. A function declaration in Python starts with the keyword **def**, followed by a bracket. Inside the bracket, the arguments to be passed to the function are specified (if there are any). Notice that the arguments are local to the function. Function **TestButton** is shown here:

```
def TestButton(Button, IN):
        if Button.value == 0:
                IN.toggle()
                while Button.value == 0;
                        pass
```

The second line of the function tests if Button is equal to 0, i.e., if the Button is pressed. Remember that normally when the Button is not pressed its state is at logic 1. If the Button is pressed, then the state of the corresponding relay pin is toggled. The function then waits until the Button is released.

```
#-----------------------------------------------------------------
#                        STATION MAINS ON-OFF CONTROL
#                        ----------------------------
# In this project 4 buttons and 4 relays are connected to Raspberry
# Pi 5.The status of the relays are toggled each time its corresonding
# button is pressed. So, if the relay is ON, it is turned OFF, if it
# is OFF it is turned ON  etc.
#
# Author: Dogan Ibrahim
# File  : relays.py
# Date  : February, 2024
#-----------------------------------------------------------------
from gpiozero import InputDevice, OutputDevice
import time

ON = 0
OFF = 1

#
# Button connections
#
BUTTONA = 4                                     # Button A
BUTTONB = 17                                    # Button B
BUTTONC = 27                                    # ButtonC
BUTTOND = 22                                    # Button D

#
# Relay connections
#
IN1 = 21                                        # IN1 pin
IN2 = 20                                        # IN2 pin
IN3 = 16                                        # IN3 pin
IN4 = 12                                        # IN4 pin

#
# IN1-IN4 are Outputs
#
in1 = OutputDevice(IN1)                         # IN1 is output
in2 = OutputDevice(IN2)                         # IN2 is output
in3 = OutputDevice(IN3)                         # IN3 is output
in4 = OutputDevice(IN4)                         # IN4 is output

#
# BUTTONA-BUTOND are inputs
#
buttona = InputDevice(BUTTONA)                  # BUTTONA is input
```

```
buttonb = InputDevice(BUTTONB)                          # BUTTONB is input
buttonc = InputDevice(BUTTONC)                          # BUTTONC is input
buttond = InputDevice(BUTTOND)                          # BUTTOND is input

#
# Turn OFF all relays at the beginning. Setting ON turns it OFF
#
in1.on()
in2.on()
in3.on()
in4.on()

#
# This function checks  if Button is pressed and toggles pin IN
# of the relay If the relay is ON, it is tured OFF, if it is OFF,
# it is turned ON
#
def TestButton(Button, IN):
    if Button.value == 0:
        IN.toggle()
        while Button.value == 0:
            pass

try:

    while True:                                         # Do forever
        TestButton(buttona, in1)                        # Test BUTTONA
        TestButton(buttonb, in2)                        # Test BUTTONB
        TestButton(buttonc, in3)                        # Test BUTTONC
        TestButton(buttond, in4)                        # Test BUTTOND
        time.sleep(0.1)

except KeyboardInterrupt:                               # Cntrl+C detected
    print("End of program")                             # End of program
```

Figure 4.8: Program: relays.py.

While developing hardware based projects on the Raspberry Pi 5 you will have to make connections to the 40-pin male type GPIO connector placed at the edge of the board. This can easily be done using a breadboard and female-male type jumper leads. One side of the jumper lead can be connected to the GPIO connector, while the other side can be connected to a breadboard so that external components can easily be interfaced to the Raspberry Pi 5. It is recommended to use a 40-way ribbon cable with a T-connector to bring all the GPIO pins to the breadboard for easy access. Figure 4.9 shows such a connector which is available on the Internet. It may however be difficult to attach a T-connector to your Raspberry Pi 5 if its fitted with a cooler.

Figure 4.9: T-connector with ribbon cable (by Sintron).

Figure 4.10 shows the project built on a breadboard and connections made using jumper wires.

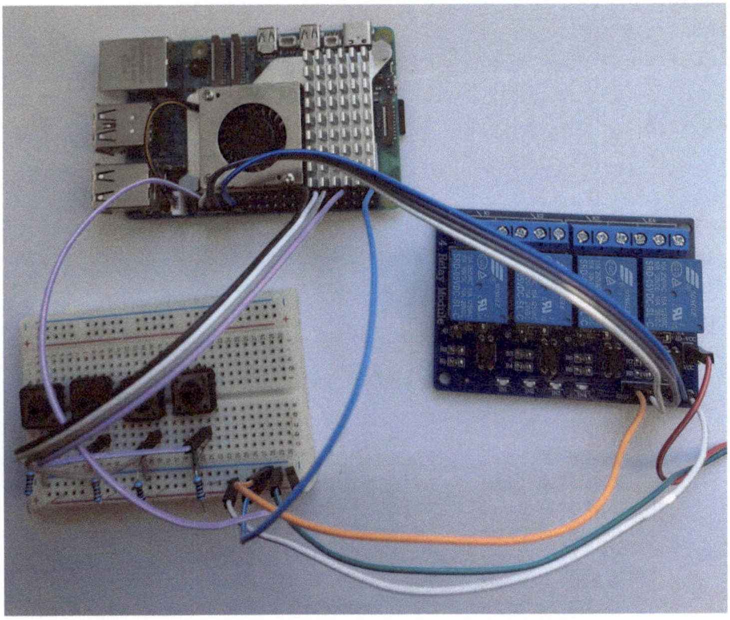

Figure 4.10: Project built on a breadboard.

4.4 Project 3: Station Clock with Output to the Monitor

Description: In this project, you will develop a real-time clock to display the current date and time on the monitor. When the Raspberry Pi 5 boots, it receives the current date and time from the Internet automatically (if it is connected to the Internet). One nice feature of Raspberry Pi 5 is that it includes an on-board Real-Time-Clock (RTC) module. If the

Internet is not available then the date and time can be read from the RTC module. An external battery can be attached to Raspberry Pi 5 to power the RTC module so that the module keeps the date and time when Raspberry Pi 5 is powered off. The RTC module can also be configured to operate with very low current and can be programmed to wake-up the processor at set intervals.

This project displays the current date and time every second on the monitor.

Program listing: Figure 4.11 shows the program listing (Program: **StnTim.py**). The program runs in an endless loop displaying the date and time as shown in Figure 4.12.

```
#=================================================================
#               CURRENT DATE AND TIME
#               =====================
#
# This program displays the current date and time on the monitor
# every second
#
# Author: Dogan Ibrahim
# File  : StnTim.py
# Date  : January, 2024
#==================================================================
import time
from datetime import date

while True:
    today = date.today()
    print("Date & Time: ",date.today().strftime("%d-%m-%Y"),time.
strftime("%H:%M:%S"))
    time.sleep(1)
```

Figure 4.11: Program listing.

```
pi@raspberrypi:~ $ python StnTim.py
Date & Time:  24-01-2024 17:24:42
Date & Time:  24-01-2024 17:24:43
Date & Time:  24-01-2024 17:24:44
Date & Time:  24-01-2024 17:24:45
Date & Time:  24-01-2024 17:24:46
Date & Time:  24-01-2024 17:24:47
Date & Time:  24-01-2024 17:24:48
Date & Time:  24-01-2024 17:24:49
Date & Time:  24-01-2024 17:24:50
Date & Time:  24-01-2024 17:24:51
Date & Time:  24-01-2024 17:24:52
Date & Time:  24-01-2024 17:24:53
```

Figure 4.12: Sample output.

4.5 Project 4: Station Clock with Output to LCD

Description: This project is similar to the previous one but the data and time are displayed on an LCD.

Block Diagram: Figure 4.13 shows the block diagram of the project. In this project an I2C type LCD is used.

Figure 4.13: Block diagram of the project.

LCDs are used as display devices in many microcontroller-based projects. Basically, there are two types of LCDs: parallel and I^2C-based. Parallel LCDs are usually controlled with 4 data lines and 2 control lines. The main advantage of these types of LCDs is their low cost. Parallel LCDs have the disadvantage that an external potentiometer is required to adjust the contrast of their screens. I^2C based LCDs are basically parallel LCDs with the addition of a small board at the back of the LCD which translates the I^2C signals to parallel signals. The advantage of the I^2C based LCDs is that their control requires only 4 wires, and the contrast adjustment potentiometer is located on the I^2C board. They are, however, more expensive than the standard parallel LCDs.

The I^2C (or I2C) bus is commonly used in microcontroller based projects. In this chapter, you will be looking at the use of this bus on the Raspberry Pi 5. Some other interesting projects are also given here. The aim is to make the reader familiar with the I^2C bus library functions and to show how they can be used in a real project. Before looking at the details of the projects, it is worthwhile to look at the basic principles of the I^2C bus.

The I^2C Bus

The I^2C bus is one of the most commonly used microcontroller communication protocols for communicating with external devices such as sensors and actuators. The I^2C bus is a single master, multiple slave bus and it can operate at standard mode: 100 Kbit/s, full speed: 400 Kbit/s, fast mode: 1 Mbit/s, and high speed: 3.2 Mbit/s. The bus consists of two open-drain wires, 'pulled up' to Vcc with resistors:

SDA: data line
SCL: clock line

Figure 4.14 shows the structure of an I^2C bus with one master and three slaves.

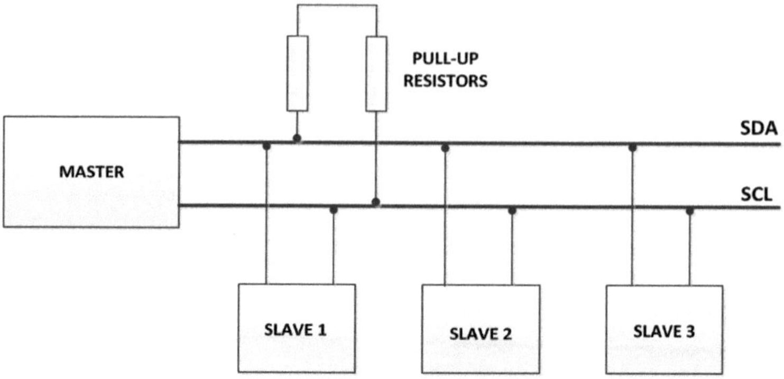

Figure 4.14: I^2C bus with one master and three slaves.

Because the I^2C bus is based on just two wires, there should be a way to address an individual slave device on the same bus. For this reason, the protocol defines that each slave device provides a unique slave address for the given bus. This address is usually 7-bits wide. When the bus is free both lines are HIGH. All communication on the bus is initiated and completed by the master, which initially sends a START bit, and completes a transaction by sending STOP bit. This alerts all the slaves that some data is coming on the bus and all the slaves listen on the bus. After the start bit, 7 bits of unique slave address is sent. Each slave device on the bus has its own address and this ensures that only the addressed slave communicates on the bus at any time to avoid any collisions. The last sent bit is read/write bit such that if this bit is 0, it means that the master wishes to write to the bus (e.g., to a register of a slave), if this bit is a 1, it means that the master wishes to read from the bus (e.g., from the register of a slave). The data is sent on the bus with the MSB bit first. An acknowledgement (ACK) bit takes place after every byte and this bit allows the receiver to signal the transmitter that the byte was received successfully and as a result another byte may be sent. The ACK bit is sent at the 9th clock pulse.

The communication over the I^2C bus is simply as follows:

- The master sends on the bus the address of the slave it wants to communicate with
- The LSB is the R/W bit which establishes the direction of data transmission, i.e., from mater to slave (R/W = 0), or from slave to master (R/W = 1)
- Required bytes are sent, each interleaved with an ACK bit, until a stop condition occurs

Depending on te type of slave device used, some transactions may require separate transaction. For example, the steps to read data from an I^2C compatible memory device are:

- Master starts the transaction in write mode (R/W = 0) by sending the slave address on the bus

- The memory location to be retrieved are then sent as two bytes (assuming 64Kbit memory)
- The master sends a STOP condition to end the transaction
- The master starts a new transaction in read mode (R/W = 1) by sending the slave address on the bus
- The master reads the data from the memory. If reading the memory in sequential format, then more than one byte will be read
- The master sets a stop condition on the bus

I²C pins of the Raspberry Pi 5

The Raspberry Pi 5 board has two I²C pins at its 40-pin GPIO header, as follows:

GPIO 2	SDA1	pin 3
GPIO 3	SCL1	pin 5
GPIO 0	SDA0	pin27
GPIO 1	SCL0	pin 28

1.8 kΩ pull-up resistors are used from the I²C pins to +3.3 V. Notice that because the I²C pins are pulled-up to +3.3 V and the Raspberry Pi 5 pins are not +5 V compatible, it is necessary to use voltage level converter circuits if the I²C LCD operates at +5 V.

The I²C LCD

The I2C LCD has 4 pins: GND, +V, SDA, and SCL. SDA can be connected to pin GPIO 2 and SCL to pin GPIO 3. +V pin of the display should be connected to the +5V (pin 2) of the Raspberry Pi 5. Raspberry Pi GPIO pins are not +5V tolerant, but the I²C LCD operates with +5V where its SDA and SCL pins are pulled to +5V. It is not a good idea to connect the LCD directly to the Raspberry Pi as it can damage its I/O circuitry. There are several solutions here. One solution is to remove the I²C pull-up resistors on the LCD module. The other option is to use an LCD which operates with +3.3V. The other solution is to use a bidirectional +3.3V to +5V logic level converter chip. In this project you will use the TXS0102 bidirectional logic level converter chip like the one shown in Figure 4.15.

Figure 4.15 Logic level converter

Note: Raspberry Pi 5 GPIO pins are claimed to be +5V tolerant as long as the RP1 module is powered ON. But for safety, a logic level converter is used in this project.

Circuit Diagram: The I²C LCD has 4 pins: GND, +V, SDA, and SCL. SDA is connected to pin GPIO2 and SCL is connected to pin GPIO3 of the Raspberry Pi respectively through a logic level converter.

Figure 4.16 shows the front and back of the I²C based LCD. Notice that the LCD has a small board mounted at its back to control the I²C interface. The LCD contrast is adjusted through the small potentiometer mounted on this board. A jumper is provided on this board to disable the backlight if required.

Figure 4.16: I²C based LCD (front and back views).

The I²C clock signal is always generated by the bus master. The devices on the I²C bus can communicate at 100 kHz or 400 kHz. Figure 4.17 shows the circuit diagram of the project.

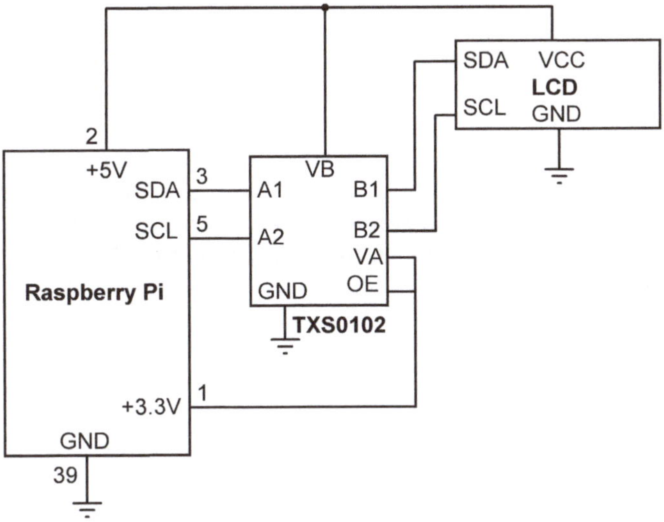

Figure 4.17: Circuit diagram of the project.

Program Listing: Before using the I²C pins of the Raspberry Pi 5, you have to enable the I²C peripheral interface on the device. The steps for this are as follows:

- Start the configuration menu from the command prompt:

  pi@raspberrypi:~ $ **sudo raspi-config**

- Go down the menu to **Interface Options**

- Go down and select **I2C**

- Enable the I²C interface

- Select **Finish** to complete

Now you have to check that the I²C library is available on your Raspberry Pi 5. The steps are as follows:

- Enter the following command. You should see the **i2c** tools (Figure 4.18:

  pi@raspberrypi:~ $ **lsmod | grep i2c**

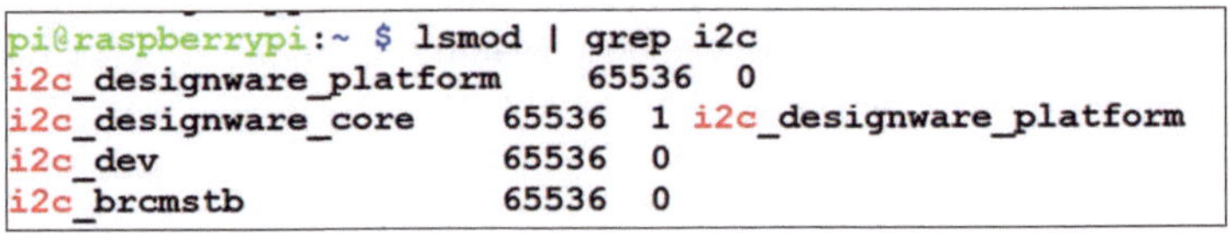

```
pi@raspberrypi:~ $ lsmod | grep i2c
i2c_designware_platform    65536  0
i2c_designware_core        65536  1 i2c_designware_platform
i2c_dev                    65536  0
i2c_brcmstb                65536  0
```

Figure 4.18: Check the I²C tools.

- Connect your LCD to the Raspberry Pi 5 device and enter the following command to check whether or not the LCD is recognized by the Raspberry Pi 5:

  pi@raspberrypi:~ $ **sudo i2cdetect –y 1**

  You should see a table similar to the one shown below. A number in the table means that the LCD has been recognized correctly and the I²C slave address of the LCD is shown in the table. In this example the LCD address is 27:

```
      0  1  2  3  4  5  6  7  8  9  a  b  c  d  e  f
00:      -- -- -- -- -- -- -- -- -- -- -- --
10:      -- -- -- -- -- -- -- -- -- -- -- -- -- -- --
20:      -- -- -- -- -- -- -- 27 -- -- -- -- -- -- --
30:      -- -- -- -- -- -- -- -- -- -- -- -- -- -- --
40:      -- -- -- -- -- -- -- -- -- -- -- -- -- -- --
50:      -- -- -- -- -- -- -- -- -- -- -- -- -- -- --
60:      -- -- -- -- -- -- -- -- -- -- -- -- -- -- --
70:      -- -- -- -- -- -- -- -- -- -- -- -- -- -- --
```

You should now install an I²C LCD library so that you can send commands and data to the LCD. There are many Python libraries available for the I²C type LCDs. The one chosen here is on GitHub from Dave Hylands. This library is installed as follows:

- Go to the following web link:

  https://github.com/dhylands/python_lcd/tree/master/lcd

- Copy the following files to your home directory **/home/pi** using WinSCP:

  i2c_lcd.py
  lcd_api.py

- Check to make sure that the files are copied successfully. You should see the files listed with the command:

  pi@raspberrypi: ~ $ **ls**

You are now ready to write the program. Figure 4.19 shows the program listing (**Stationtim.py**). At the beginning of the program the LCD driver libraries **lcd_api** and **i2c_lcd** are imported to the program. The heading DATE AND TIME is displayed for three seconds at the top row (row 1) and the program enters a loop. Inside this loop the current date and time are displayed row 1 and row 2 of the LCD respectively.

```
#-----------------------------------------------------------------
#                        STATION DATE AND TIME
#                        ---------------------
# In this project an I2C LCD is connected to the Raspberry Pi 5.
# The program gets the current date and time and displays on
# the LCD
#
# Author: Dogan Ibrahim
# File  : Stationtim.py
# Date  : February, 2024
#-----------------------------------------------------------------
import time
from datetime import date

#
# Import LCD libraries
#
from lcd_api import LcdApi
from i2c_lcd import I2cLcd

I2C_ADDR = 0x27
I2C_NUM_ROWS = 2
I2C_NUM_COLS = 16

mylcd = I2cLcd(1,I2C_ADDR,I2C_NUM_ROWS,I2C_NUM_COLS)

mylcd.clear()                                   # clear LCD
mylcd.putstr("DATE AND TIME")                   # display heading
time.sleep(3)                                   # Wait 3 seconds

#
# Get current date and time. Display date at first row and time
# in the second row of the LCD
#
try:

  while True:
      mylcd.clear()
      today = date.today()
      mylcd.move_to(0,0)
      mylcd.putstr(date.today().strftime("%d-%m-%Y"))
      mylcd.move_to(0,1)
      mylcd.putstr(time.strftime("%H:%M:%S"))
      time.sleep(1)

except KeyboardInterrupt:                       # Cntrl+C detected
  print("End of program")                       # End of program
```

Figure 4.19: Program listing.

Figure 4.20 shows example output on the LCD.

Figure 4.20: Example output on LCD.

The I²C LCD library supports many functions. Some of the commonly used functions are (see the LCD library documentation for more details):

clear()	clear LCD and set to home position
show_cursor()	show cursor
hide_cursor()	hide cursor
blink_cursor_on()	blink cursor
blink_cursor_off()	stop blinking cursor
display_on()	display on
display_off()	display off
backlight_on()	backlight on
backlight_off()	backlight off
move_to(x, y)	move cursor to (x, y)
putchar()	display a character
putstr()	display a string

4.6 Project 5: Station Geographical Coordinates

There are requirements, especially when working mobile that we may want to know the geographical coordinates (e.g., latitude and longitude) of our station. In this project, we use a GPS to read and display the geographical coordinates of our station on an LCD.

Description: GPS receivers receive geographical data from the GPS satellites and they provide accurate information about the position of the user on Earth. These satellites circle the Earth at an altitude of about 20,000 kms and complete two full orbits every day. In order for a receiver to determine its position, it is necessary that for the receiver to communicate with at least 3 satellites. Therefore, if the receiver does not have clear view of the sky then it may not be possible to determine its position on Earth. In some applications external antennas are used so that even very weak signals can be received from the GPS satellites.

The data sent out from a GPS receiver is in text format and is known as the NMEA Sentences. Each NMEA sentence starts with a $ character and the values in a sentence are separated by commas. Some of the NMEA sentences returned by a GPS receiver are given below:

3$GPGLL: This sentence returns the local geographical latitude and longitude.

$GPRMC: This sentence returns the local geographical latitude and longitude, speed, track angle, date, time, and magnetic variation.

$GPVTG: This sentence true track, magnetic track, and the ground speed.

$GGGA: This sentence returns the local geographical latitude and longitude, time, fix quality, number of satellites being tracked, horizontal dilution of position, altitude, height of geoid, and DGPS data.

$GPGSV: There are 4 sentences with this heading. These sentences return the number of satellites in view, satellite number, elevation, azimuth, and SNR.

In this project, the GPS Click board (www.mikroe.com) is used to receive the NMEA sentences. This is a small GPS receiver (see Figure 4.21) which is based on the LEA-6S type GPS. This board operates with +3.3 V and provides two types of outputs: I²C or serial output. In this project the default serial output is used which operates at 9600 Baud rate. An external dynamic antenna can be attached to the board in order to improve its reception for indoor use or in for use in places where there may not be clear view of the sky.

Figure 4.21: GPS Click board.

Figure 4.22 shows the complete list of the NMEA sentences output from the GPS Click board every second.

```
$GPGLL,5127.3917,N,00003.13141,E,10534.00,A,A*67
$GPRMC,05305.00,A,5127.35909,,0003.13148,E,0.030,,270919,,,A*7E
$GPVTG,,T,,M,0030,N,0.055,K,A*20
$GGGA,105305.00,5127.35909,N,00003.13148,E,1,09,1.18,46.5,M,45.4,M,,*66
$GPSA,A,3,01,32,08,28,18,03,22,14,11,,,,2.12,1.18,1.76*06
$GPGSV,4,1,13,01,7,304,40,03,40,224,31,08,38,165,32,10,05,054,*77
$GPGSV,4,2,13,11,83,217,3,14,39,094,24,17,17,314,22,18,73,091,41*76
$GPGSV,4,3,13,22,63,219,33,24,1,002,,27,05,150,,28,30,284,28*7F
$GPGSV,4,4,13,32,34,063,35*4E
```

Figure 4.22: NMEA sentences output from the GPS Click board.

GPS Click board is a 2x8 pin dual-in-line module and it has the following pin configuration (pin 1 is the top left pin of the module):

1: No connection	16: No connection
2: Reset	15: No connection
3: No connection	14: TX
4: No connection	13: RX
5: No connection	12: SCL

6: No connection	11: SDA
7: +3.3V	10: No connection
8: GND	9: GND

In serial operation, only the following pins are required: +3.3V, GND, TX. In this project, an external dynamic antenna is attached to the GPS Click board as it was tested indoors.

$GPGLL is one of the commonly used NMEA sentence and this is the sentence used in this project to extract the station geographical coordinates. This sentence is output every second in the following format:

$GPGLL,5127.37032,N,00003.12782,E,221918.00,A,A*61

The fields in this sentence can be decoded as follows:

GLL	Geographic position, latitude and longitude
5127.37032	Latitude 51 deg., 27.3702 min. North
00003.12782	Longitude 0 deg., 3.12782 min. East
221918	Fix taken at 22L19L18 UTC
A	Data active (or V for void)
*61	checksum data

Notice that the received data fields are separated by commas. The validity of the received data is shown by letters **A** or **V** in the data, where A shows that the data is valid, and V indicates that the data is not valid. A checksum is used at the end of the data.

Block Diagram: Figure 4.23 shows the block diagram of the project. An I^{2}C LCD is used to display the latitude and longitude of the station.

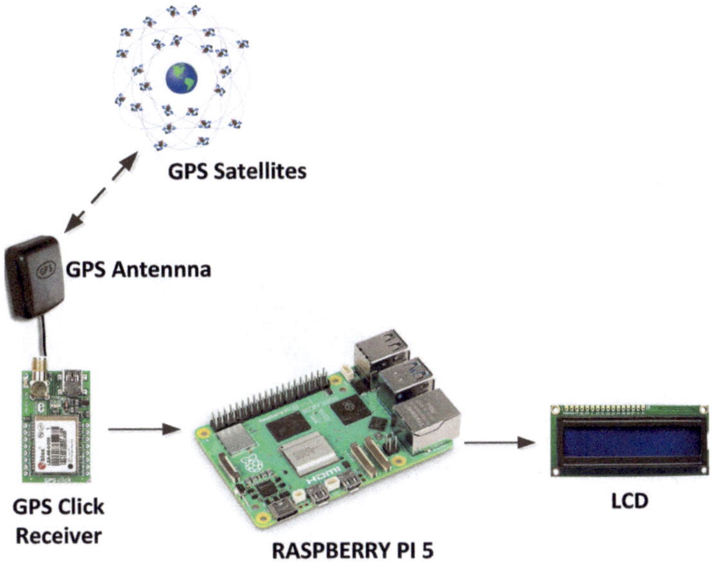

Figure 4.23: Block diagram of the project.

The *Serial Debug* port, marked as **UART** on the board (Figure 4.24), is used to communicate with the GPS click board. This serial port is identified as **/dev/ttyAMA10**.

Figure 4.24: Raspberry Pi 5 Serial Debug port (UART)

The Serial Debug port is interfaced to external world through a 3-pin JST-SH type micro connector and cables (Figure 4.25).

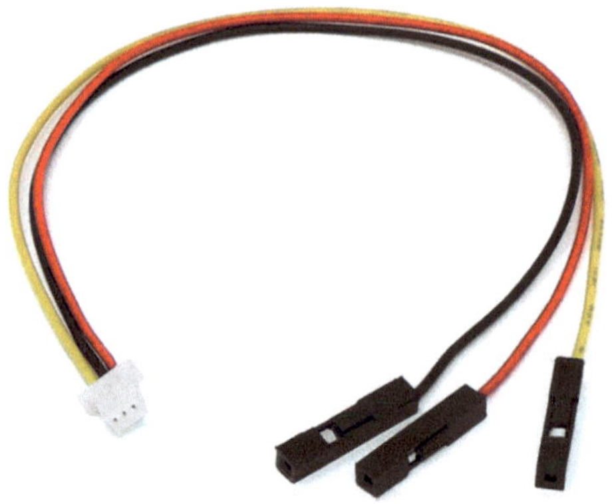

Figure 4.25: 3-pin JST-SH connector and cable.

The JST-SH cable is identified with different colors:

Orange:	RX (serial input to Raspberry Pi 5)
Black:	Ground
Yellow:	RX (serial output from Raspberry Pi 5)

Circuit Diagram: The circuit diagram of the project is shown in Figure 4.26. The UART TX pin of the GPS click board (pin 14) and Ground (pin 8) are connected to RX (yellow) and Ground (black) pins of the cable. The I^2C LCD is connected as in the previous projects. i.e., the SDA and SCL pins are connected to GPIO2 and GPIO3 respectively through a logic level converter module. The GPS click board is powered from the Raspberry Pi 5's +3.3 V supply.

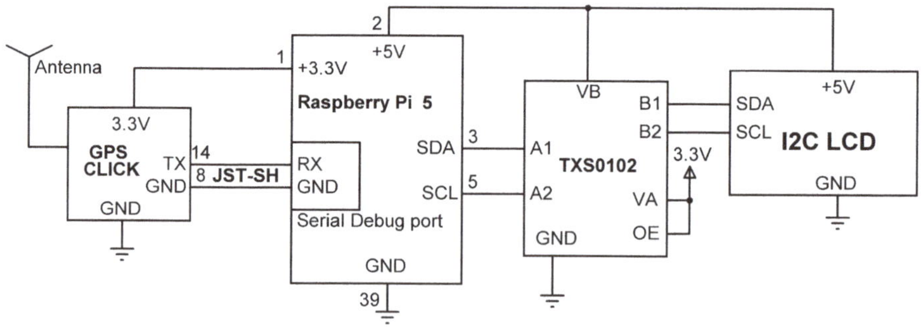

Figure 4.26: Circuit diagram of the project.

We can display all the GPS NMEA sentences sent by the GPS click board by entering the following command. This command is useful for testing purposes since it verifies if the connection to the GPS is correct and also if the GPS is receiving data from the satellites :

pi@raspberrypi:~ $ **sudo cat /dev/ttyAMA10**

An example output is shown in Figure 4.27 (only part of the output is shown).

```
pi@raspberrypi:~ $ sudo cat /dev/ttyAMA10
$GPRMC,100129.00,A,5127.37042,N,00003.12901,E,0.278,,270124,,,A*7F

$GPVTG,,T,,M,0.278,N,0.515,K,A*2F

$GPGGA,100129.00,5127.37042,N,00003.12901,E,1,06,1.19,57.1,M,45.4,M,,*65

$GPGSA,A,3,17,15,10,24,12,14,,,,,,,2.26,1.19,1.92*04

$GPGSV,2,1,08,02,04,012,20,10,26,306,41,12,31,210,27,14,12,050,29*7F

$GPGSV,2,2,08,15,52,175,25,17,30,065,15,24,80,289,28,25,01,223,*76

$GPGLL,5127.37042,N,00003.12901,E,100129.00,A,A*69
```

Figure 4.27: Example output of NMEA sentences.

In this project, the latitude and longitude are extracted from the NMEA sentence $GPGLL without using a library.

Program Listing: Figure 4.28 shows the program listing (program: **gps.py**). Variable **port** is assigned to **/dev/ttyAMA10** which is the debug serial port name for Raspberry Pi 5. Function **Get_GPS()** receives a line of NMEA sentence and looks for string **$GPGLL**. When this string is detected the line of sentence is broken down into parts separated by commas using built-in function **split(",")** and stored in **sdata**. If the 6th field is character **V** then it is assumed that the sentence is not valid (e.g., there is no satellite reception) and the text **NO DATA** is displayed at the top row of the LCD. Otherwise, the latitude and its direction are extracted from fields 1 and 2 and stored in variables **lat** and **latdir** respectively. The longitude and its direction are extracted from fields 3 and 4 and stored in variables **lon** and **londir** respectively.

The latitude is received in the format: **ddmm.mmmmmD** which corresponds to **dd** degrees **mm.mmmmm** minutes, and direction **D** which is **N** or **S**. Similarly, the longitude is received in the format: **dddmm.mmmmmD** where **D** is **E** or **W**. The main program separates the degrees and minutes and displays them on the LCD. The latitude is displayed at the top row of the LCD in the format: **dd mm.mmmmm D**, and the longitude is displayed as: **ddd mm.mmmmm D**.

```
#--------------------------------------------------------------
#               STATION GEOGRAPHICAL COORDINATES
#               --------------------------------
#
# In this project a GPS receiver module (GPS CLICK) is connected
# to the serial debug port of the Raspberry Pi 5. Additionally,
# an I2C LCD is connected. The program displays the latitude and
# longitude of the receiver location on the LCD
#
# Author: Dogan Ibrahim
# File  : gps.py
# Date  : January, 2024
#--------------------------------------------------------------
import time
```

```
#
# Import LCD libraries
#
from lcd_api import LcdApi
from i2c_lcd import I2cLcd

#
# LCD configuration
#
I2C_ADDR = 0x27
I2C_NUM_ROWS = 2
I2C_NUM_COLS = 16

mylcd = I2cLcd(1,I2C_ADDR,I2C_NUM_ROWS,I2C_NUM_COLS)
mylcd.clear()                              # clear LCD

import serial                              # Import srial

port = "/dev/ttyAMA10"                     # Serial port
lat=latdir=lon=londir = "0"

#
# This function receives and extracts the latitude and longitude
# from the NME sentence $PGLL
#
def Get_GPS(data):
   global lat,latdir,lon,londir
   dat = data.decode('utf-8')
   if dat[0:6] == "$GPGLL":
       sdata = dat.split(",")                  # SPlit data
       if sdata[6] == "V":                     # Valid data?
           mylcd.clear()                       # Clear LCD
           mylcd.move_to(0, 0)                 # At 0,0
           mylcd.putstr("NO DATA")             # No data
           return
       lat = sdata[1]                          # Get latitude
       latdir = sdata[2]                       # Latitude dir
       lon = sdata[3]                          # Get longitude
       londir = sdata[4]                       # Longitude dir
       return

#
# Receive the GPS coordinates and display on the LCD
#
ser = serial.Serial(port,baudrate=9600,timeout=0.5)
```

```
try:

   while True:
      data = ser.readline()                          # Read a line
      Get_GPS(data)                                  # Decode

      deg = lat[0:2]
      min = lat[2:]
      latitude = str(deg) + " " + str(min) + " " + str(latdir)

      deg = lon[0:3]
      min = lon[3:]
      longitude = str(deg) + " " + str(min) + " " + str(londir)

      mylcd.clear()
      mylcd.move_to(0, 0)
      mylcd.putstr(latitude)                         # Display latitude
      mylcd.move_to(0, 1)
      mylcd.putstr(longitude)                        # Display longitude
      time.sleep(1)                                  # WAit 1 secons

except KeyboardInterrupt:                            # Cntrl+C detected
   ser.close()                                       # Close serial
   print("End of program")                           # End of program
```

Figure 4.28: Program listing.

An example display on the LCD is shown in Figure 4.29.

Figure 4.29: Example display on the LCD.

Figure 4.30 shows the project built on a breadboard and connections made using jumper wires.

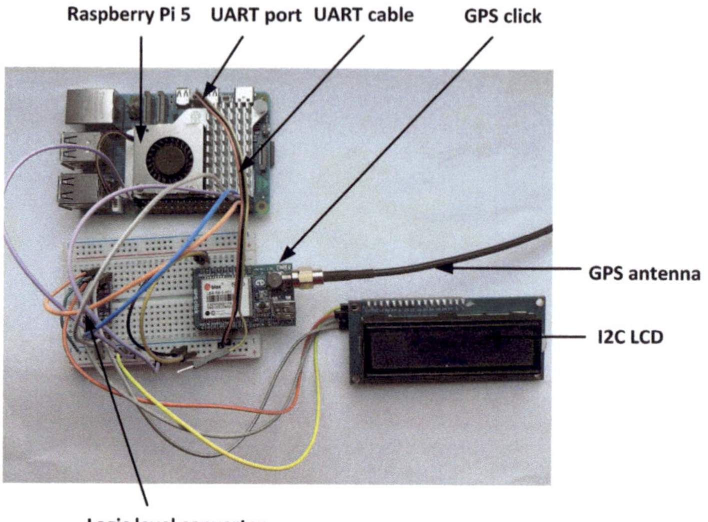

Figure 4.30: Project built on a breadboard.

4.7 Project 6: Waveform Generation in Software — Sawtooth waveform

Waveform generators are important tools for amateur radio operators. There are many commercially available equipment such as signal generators for generating various waveforms. The author's earlier book entitled *Raspberry Pi for Radio Amateurs* gives examples of several projects for generating square wave, triangular wave, sawtooth wave, sine wave etc using a Raspberry Pi 4. In this section, we will give an example of generating a sawtooth wave using a Raspberry Pi 5. Readers who have the earlier book and Raspberry Pi 5 should be able to convert the programs in that book to work on the Raspberry Pi 5.

Before generating an analog waveform, it is necessary to use a Digital-to-Analog Converter chip (DAC) to convert the generated digital signals into analog form. In this book we will be using the popular MCP4921 DAC chip from Microchip.

4.7.1 The MCP4921 DAC

Before using the MCP4921, it is worthwhile to look at its features and operation in some detail. MCP4921 is a 12-bit DAC that operates with the SPI bus interface. Figure 4.31 shows the pin layout of this chip. The basic features are:

- 12-bit operation
- 20 MHz clock support
- 4.5 µs settling time
- External voltage reference input
- Unity or 2× gain control
- 1× or 2× gain
- 2.7 to 5.5 V supply voltage
- –40 ºC to +125 ºC temperature range

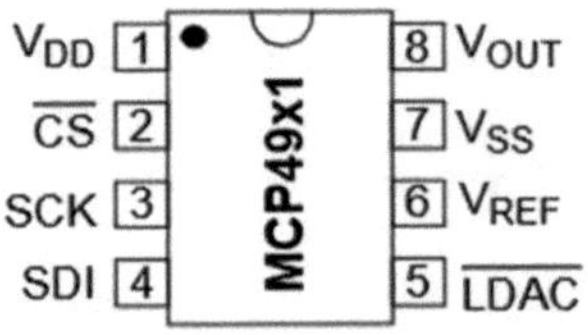

Figure 4.31: The MCP4921 DAC.

The pin descriptions are:

Vdd:	supply voltage
CS:	chip select (active LOW)
SCK:	SPI clock
SDI:	SPI data in
LDAC:	Used to transfer input register data to the output (active LOW)
Vref	Reference input voltage
Vout:	analog output
Vss:	supply ground

In this project, you will be operating the MCP4921 set to a gain of 1. As a result, with a reference voltage of 3.3 V and 12-bit conversion data, the LSB resolution of the DAC will be 3300 mV / 4096 = 0.8 mV.

The SPI Bus

The Serial Peripheral Interface (SPI) bus consists of two data wires and one clock wire. Additionally, a chip enable (CE or CS) connection is used to select a slave in a multi-slave system. The wires used are:

MOSI (or **SDI)**: **M**aster **O**ut **S**lave **I**n. This signal is output from the master and is input to a slave.

MISO: **M**aster **I**n **S**lave **O**ut. This signal is output from a slave and input to a master.

SCLK (or **SCK)**: The clock, controlled by the master.

CE (or **CS)**: **C**hip **E**nable (slave select).

There are four SPI operational modes known as mode 0 to mode3. The mode determines the relationship between the clock pulses and the data pulses. Data can be read at the leading edge or at the trailing edge of the clock. CPOL (clock polarity) and CPHA (clock phase) determine the mode of operation. CPOL determines the polarity of the clock. In mode 0, both CPOL and CPHA are 0 and data is sampled at the leading rising edge of the clock. In mode 2, CPOL=1 and CPHA=0, data is sampled at the leading falling edge of the clock. In mode 1, CPOL=0 and CPHA=1 and the data is sampled on the trailing falling edge of the clock. Finally, in mode 3, CPOL=1 and CPHA=1 where the data is sampled on the trailing rising edge of the clock. Mode 0,0 is used by the Raspberry Pi which is the most commonly used mode.

The following pins are the SPI bus pins on Raspberry Pi 5:

GPIO pin	**SPI**	**Physical pin no**
GPIO10	MOSI	19
GPIO9	MISO	21
GPIO11	SCLK	23
GPIO8	24	CE0
GPIO7	26	CE1

The SPI bus must be enabled on the Raspberry Pi 5 before it can be used. The steps are:

- Start the configuration tool:

  pi@raspberrypi:~ $ **sudo raspi-config**

- Select **Interfacing Options**.
- Select **SPI**.
- Select **Yes** to enable it.
- Select **Finish**.
- Select to reboot your Raspberry Pi 5.

After the system comes up, enter the following command to confirm that the SPI bus has been enabled:

pi@raspberrypi:~ $ **ls /dev/*spi***

You should get a response similar to:

/dev/spidev0.0 /dev/spidev0.1

This represents that there could be up to two SPI devices on chip enable pins 0 and 1. Note that you can have more SPI devices daisy-chained and sharing the same chip enable pin if you want.

The SPI bus on Raspberry Pi 5 supports the following functions:

Function	**Description**
open (0,0)	Open SPI bus 0 using CE0
open (0,1)	Open SPI bus 0 using CE1
close()	disconnect the device from the SPI bus
writebytes([array of bytes])	Write an array of bytes to SPI bus device
readbytes(len)	Read *len* bytes from SPI bus device

xfer2([array of bytes])	Send an array of bytes to the device with CEx asserted at all times
xfer([array of bytes])	Send an array of bytes de-asserting and asserting CEx with every byte transmitted

Description: In this project we will be using the MCP4921 DAC to generate a sawtooth wave signal with the following specifications:

Peak voltage:	3.3 V
Step width:	1 ms
Number of steps:	6

Block diagram: Figure 4.32 shows the block diagram of the project. The output voltage is plotted using a Velleman PCSGU250 model oscilloscope & function generator. Pin GPIO26 is used as an output and is connected to the oscilloscope.

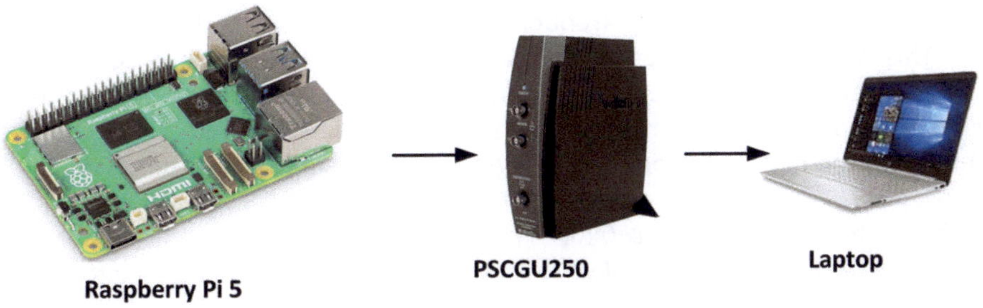

Figure 4.32: Block diagram of the project.

Circuit Diagram: The circuit diagram of the project is as shown in Figure 4.33

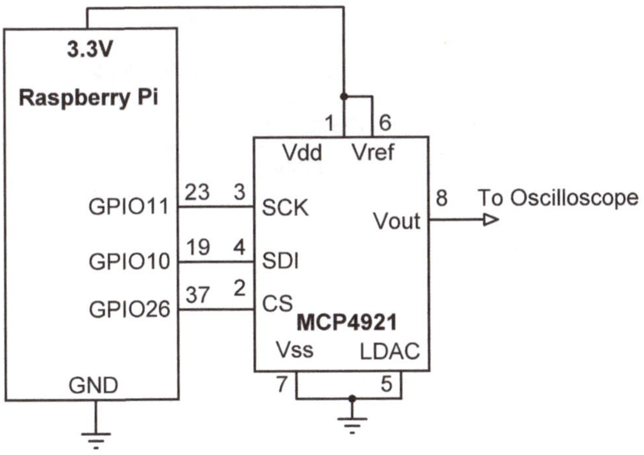

Figure 4.33: Circuit diagram of the project.

Program Listing: Data is written to the DAC in 2 bytes. The lower byte specifies D0:D8 of the digital input data. The upper byte consists of the following bits:

D8:D11	bits D8:D11 of the digital input data
SHDN	1: active (output available), 0: shutdown the device
GA	output gain control. 0: gain is 2×, 1: gain is 1×
BUF	0: input unbuffered, 1: input buffered
A/B	0: write to DACa, 1: Write to DACb (MCP4921 supports only DACa)

In normal operation, you send the upper byte (D8:D11) of the 12 bit (D0:D11) input data with bits D12 and D13 set to 1 so that the device is active and the gain is set to 1x. Then you send the low byte (D0:D7) of the data. This means that 0x30 should be added to the upper byte before sending it to the DAC.

Figure 4.34 shows the program listing (program: **sawtooth.py**). At the beginning of the program GPIO26 is used as the **CS** pin and is configured as an output. Variable **frequency** is set to 1000 which is the required frequency. Function DAC sends the 12 bit input data to the DAC. This function has two parts. In the first part, the HIGH byte is sent after adding 0x30 as described above. Function **xfer2** is used to send the data to the DAC. In the second part of the function, the LOW byte is extracted and is sent to the DAC. Notice that we could have send both the high byte and the low byte using the same **xfer2** function as follows:

```
highbyte = (data >> 8) & 0x0F
highbyte = highbyte + 0x30

lowbyte = data & 0xFF
xfer2([highbyte, lowbyte])
```

Notice that the speed of the SPI interface is set to 3900000 which corresponds to 3.9MHz. The table below shows the values set for the speed and the actual speed of the SPI interface:

Speed	spi.max_speed_hz value
125.0 MHz	125000000
62.5 MHz	62500000
31.2 MHz	31200000
15.6 MHz	15600000
7.8 MHz	7800000
3.9 MHz	3900000
1953 kHz	1953000
976 kHz	976000
488 kHz	488000
244 kHz	244000

122 kHz	122000
61 kHz	61000
30.5 kHz	30500
15.2 kHz	15200
7629 Hz	7629

```
#--------------------------------------------------------------------
#                GENERATE SAWTOOTH WAVEFORM
#                ==========================
#
# This program generates sawtooth waveform with 6 steps where each
# step has a width of 1ms
#
# Author: Dogan Ibrahim
# File  : sawtooth.py
# Date  : February, 2024
#----------------------------------------------------------------------
from gpiozero import OutputDevice
import time                                  # Import time
import spidev                                # Import SPI

spi = spidev.SpiDev()
spi.open(0, 0)                               # Bus=0, device=0
spi.max_speed_hz = 3900000

CS = 26                                      # GPIO26 is CS output
cs = OutputDevice(CS)

cs.on()                                      # Disable CS

#
# This function implements the DAC. The data in "data" is sent
# to the DAC
#
def DAC(data):
   cs.off()                                  # Enable CS

#
# Send HIGH byte
#
   temp = (data >> 8) & 0x0F                 # Get upper byte
   temp = temp + 0x30                        # OR with 0x30
   spi.xfer2([temp])                         # Send to DAC
#
# Send LOW byte
```

```
#
   temp = data & 0xFF                          # Get lower byte
   spi.xfer2([temp])                           # Send to DAC

   cs.on()                                     # Disable CS

try:

  while True:                                  # Do forever
     i = 0
     while i < 1.1:
        DACValue = int(i*4095)                 # Value  to send
        DAC(DACValue)                          # Send to DAC
        time.sleep(0.0007)                     # Wait
        i = i + 0.2

except KeyboardInterrupt:
  exit(0)
```

Figure 4.34: Program listing.

An example output waveform taken from the oscilloscope is shown in Figure 4.35. Notice that the time delay had to be adjusted experimentally to give the correct timing.

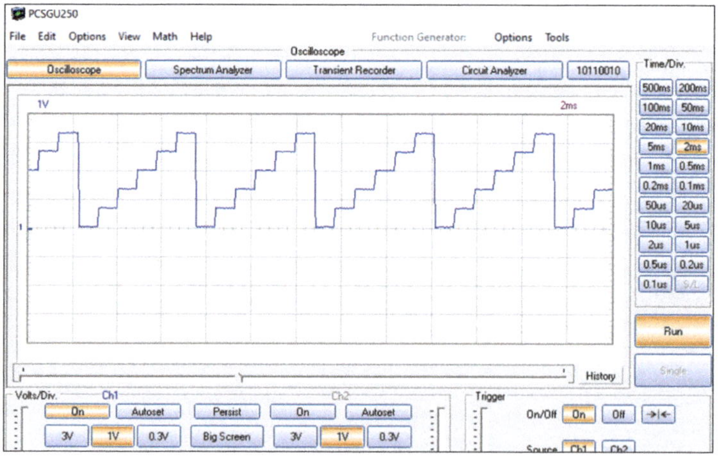

Figure 4.35: Example output waveform.

4.8 Project 7: Generating a Waveform – Frequency Entry using Keypad and LCD

In this section, you will explore the use of a programmable signal generator module that can be used to generate accurate sine and square wave signals. You will be using the popular AD9850 signal generator module together with the Raspberry Pi 5.

4.8.1 The AD9850

The AD9850 (see Figure 4.36) module is a dual sine- and square-wave generator with the following features:

- 0 – 40 MHz sine wave output.
- 0 – 1 MHz square wave output.
- 3.3 V or 5 V operation.
- 125 MHz on-board timing crystal.
- On-chip 10 bit DAC.
- Serial or parallel programming data.
- 60 mA operating current.

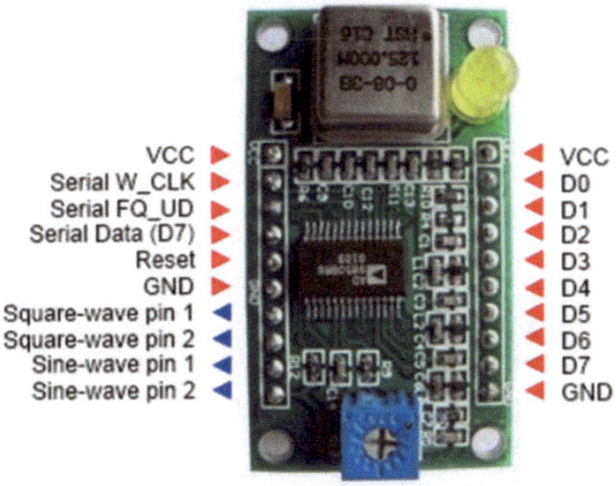

Figure 4.36: The AD9850 module.

The AD9850's circuit architecture allows the generation of output frequencies of up to one-half the reference clock frequency (or 62.5 MHz). The device also provides five bits of digitally controlled phase modulation, which enables phase shifting of its output in increments of 180°, 90°, 45°, 22.5°,11.25°, and any combination thereof.

Before using the AD9850 in a project, you need to know how to program it, and this is described briefly in the following sections (more information on the AD9850 can be obtained from the manufacturers' data sheet at:

https://www.analog.com/media/en/technical-documentation/data-sheets/ad9850.pdf)

Figure 4.37 shows the AD9850 pin layout.

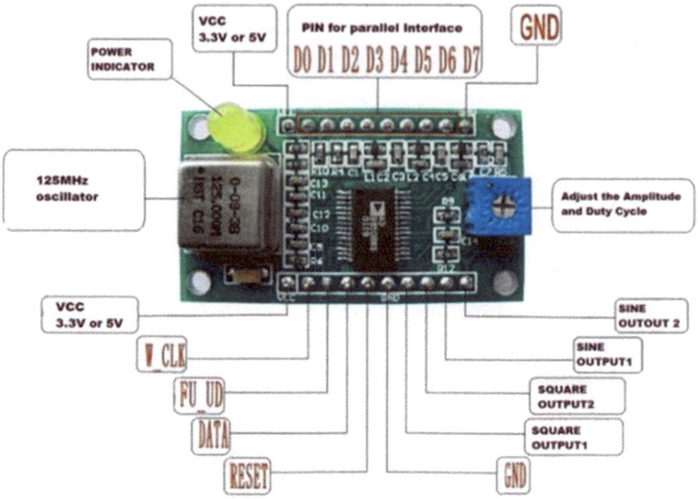

Figure 4.37: AD9850 module pin layout.

The pin definitions are:

VCC: supply voltage (3.3 V or 5 V)
GND: supply ground
W_CLK: load word clock (used to load parallel or serial frequency/phase words)
FQ_UD: frequency update (the frequency or phase is updated on rising edge of this pin)
DATA: serial load pin
RESET: master reset pin (HIGH to reset to clear all registers)
D0-D7: parallel data input (for loading 32-bit frequency and 8-bit phase/control word)
Square Wave Output1: square wave output 1 (comparator output)
Square Wave Output2: square wave output 2 (comparator complement output)
Sine Wave Output1: analog sine wave output 1 (DAC output)
Sine Wave Output2: analog sine wave output 2 (DAC complement output)

The on-board potentiometer is used to set the duty cycle of the square waveform.

The AD9850 can be loaded either in parallel or in serial form. In this project, you will be using the **serial mode**. The relationship between the output frequency, reference clock, and tuning word of the AD9850 is determined by the following formula:

$$F_{out} = \Delta\text{Phase} \times \text{REFCLOCK} / 2^{32}$$

or

$$\Delta\text{Phase} = F_{out} \times 2^{32} / \text{REFCLOCK}$$

where F_{out} is the output frequency in MHz, ΔPhase is the value of the 32-bit tuning word, and REFCLOCK is the reference clock in MHz. The output sine wave is an analog signal output by a 10-bit DAC. Notice that 2^{32} = 4,294,967,296.

The AD9850 contains a 40-bit register, consisting of the 32-bit frequency control word, 5-bit phase modulation word, and the power-down function. In this project this register is loaded serially. In this mode, rising edge of **W_CLK** shifts the 1-bit data on pin D7 (DATA) through the 40-bits of programming information. After shifting 40-bits, an **FQ_UD** pulse is required to update the output frequency (or phase).

The serial mode is enabled by the following sequence (see Data Sheet page 12, Figure 10):

- Pulse RESET
- Pulse W_CLK
- Pulse FQ_UD

The 40-bit serial data is loaded as follows:

- Send bit 0 - frequency (LSB)
- Send bit 1 - frequency
- Send bit 2 - frequency
- ....
- ....
- Send bit 31 - frequency (MSB)
- Send bit 32 – Control (set to 0)
- Send bit 33 – Control (set to 0)
- Send bit 34 - power down
- Send bit 35 - phase (LSB)
- Send bit 36 - phase
- Send bit 37 – phase
- Send bit 38 – phase
- Send bit 38 – phase
- Send bit 39 – phase (MSB)

Description: In this project a keypad and an I^2C LCD are used to set the frequency of the waveform. This makes the project autonomous so that for example the Raspberry Pi 5 can be used on its own without having to use a PC to set the frequency of the waveform. In this project the generated sine wave frequency is set to 2000 Hz.

Block Diagram: Figure 4.38 shows the block diagram of the project. The keypad sets the desired frequency which is displayed on the LCD as it is entered. The generated signal waveform is output by the AD9850 module which can be displayed on an oscilloscope if desired.

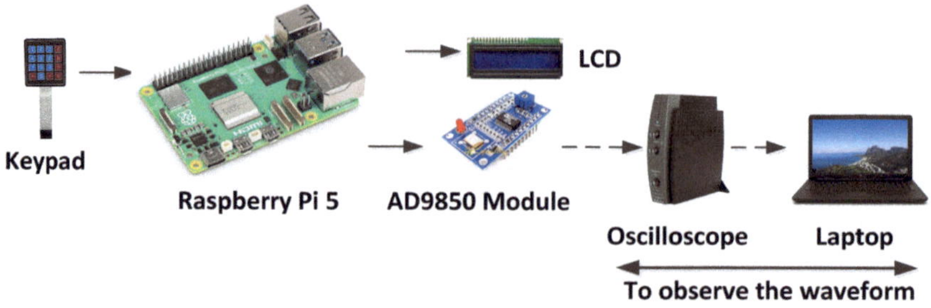

Figure 4.38: Block diagram of the project.

The Keypad: There are several types of keypads that can be used in microcontroller based projects. In this project a 4×4 keypad (see Figure 4.39) is used. This keypad has keys for numbers 0 to 9 and letters A, B, C, D, *, and #. The keypad is interfaced to the processor with 8 wires with the names R1 to R4 and C1 to C4, representing the rows and columns respectively of the keypad (see Figure 4.40).

Figure 4.39: 4×4 keypad used in the project.

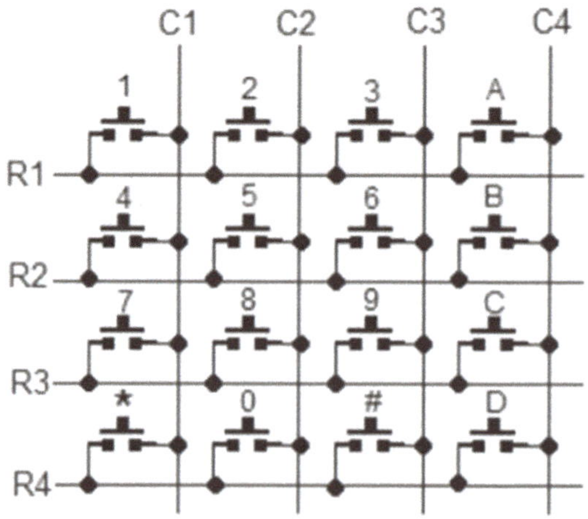

Figure 4.40: Circuit diagram of the 4×4 keypad.

The operation of the keypad is very simple: the columns are configured as outputs and the rows as inputs. The pressed key is identified by using column scanning. Here, a column is forced low while the other columns are held high. Then the state of each row is scanned, and if a row is found to be low, then the key at the intersection of the row which is low and this column is the key pressed. This process is repeated for all the rows.

Circuit Diagram: Figure 4.41 shows the circuit diagram of the project. The I^2C LCD is connected to the Raspberry Pi 5 as in the previous projects, where GPIO 2 and GPIO 3 are used as the SDA and SCL pins respectively. The AD9850 module and the 4x4 keypad are connected to the following GPIO pins of the Raspberry Pi 5.

AD9850 module pin	**Raspberry Pi 5 pin**
FQ_UD	GPIO26
W_CLK	GPIO19
RESET	GPIO13
DATA	GPIO6

Keypad pin	**Raspberry Pi 5 pin**
R1	PIO 14
R2	GPIO 15
R3	GPIO 12
R4	GPIO 23
C1	GPIO 24
C2	GPIO 25
C3	GPIO 8
C4	GPIO 7

The keypad row pins are held high using 10-kΩ pull-up resistors to +3.3 V. **Notice** that the Raspberry Pi 5 has internal pull-up resistors when a pin is used as an input but the use of these pull-up resistors is not reliable.

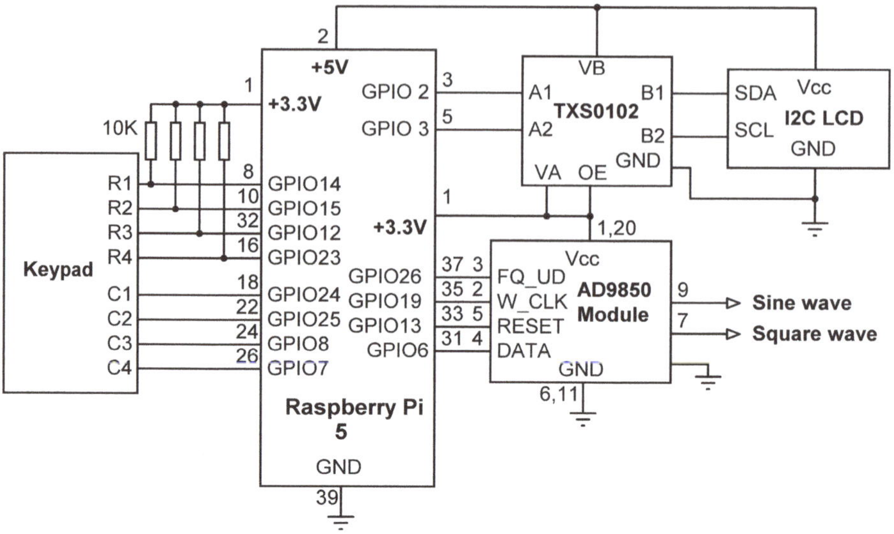

Figure 4.41: Circuit diagram of the project.

Figure 4.42 shows the pin configuration of the 4×4 keypad used in the project.

Figure 4.42: Pin configuration of the 4x4 keypad.

Program Listing: The SPI interface must be disabled using the **raspi-config** command before the program is run. This is because some of the ports used by the program are shared with the SPI interface and this gives GPIO busy errors.

You are now ready to develop the program of this project. Figure 4.43 shows the program listing (Program: **keypadsig.py**). At the beginning of the program the connections of

AD9850 pins **FQ_UD**, **W_CLK**, **RESET**, and **DATA** are set to 26, 19, 13, and 6 respectively which are the GPIO pin names. These pins are configured as outputs and all are cleared to 0 at the beginning of the program. Also, connections to the keypad and the LCD are defined. The program consists of a number of functions:

SendPulse: This function sends a HIGH-LOW logic signals to the pin specified in its argument GPIOpin.

SendByte: This function receives a byte as its argument. Then a loop is formed to send the 8 bits of this byte to serial input DATA. A pulse is sent to W_CLK after sending a bit.

SetSerialMode: This function puts the AD9850 into serial mode by pulsing pins RESET, W_CLK, and FQ_UD.

LoadFrequency: This function receives the required frequency as its argument. Then the tuning word is calculated using the formula described earlier in this section and is stored in variable **f.** Then a for loop is formed to iterate 4 times. The LOW byte of **f** is then sent to function **SendByte** so that its 8 bits are sent to the serial input. Variable **f** is then shifted right 8 times so that the next higher byte is serialized and sent to function **SendByte**. This process is repeated for the 4 bytes, making a total of 32-bits. Then, the remaining 8 bits are sent as zeroes so that the two Control words are 0, and also the phase bits are 0.

In this program, key **D** is assumed to be the **ENTER** key where all the inputs to the keypad must be terminated by pressing the **ENTER** key. Only keys 0 to 9 and key **D** are accepted by the program. Pressing any other key is ignored by the program. When key **D** is pressed, the program stores the entered frequency in variable **Total** which is then copied to variable **frequency**. After pressing the **D** key, the LCD displays **OK** to confirm that the entered frequency value is accepted by the program. After a short delay, the second row of the LCD is cleared so that it is ready to accept a new frequency value.

```
#--------------------------------------------------------------
#              FREQUENCY GENERATOR WITH KEYPAD AND LCD
#              ======================================
#
# In this program the required frequency is entered from a keypad
# in Hz. An LCD is used to display the entered frequency
#
# Author: Dogan Ibrahim
# File  : keypadsig.py
# Date  : February, 2024
#--------------------------------------------------------------
import time
from gpiozero import OutputDevice, InputDevice
#
# Import LCD libraries
#
```

```
from lcd_api import LcdApi
from i2c_lcd import I2cLcd

I2C_ADDR = 0x27
I2C_NUM_ROWS = 2
I2C_NUM_COLS = 16

mylcd = I2cLcd(1,I2C_ADDR,I2C_NUM_ROWS,I2C_NUM_COLS)
mylcd.clear()                                 # clear LCD

#
# Keypad connections
#
KEYPAD = [                                    # Keypad keys
    [1,2,3,"A"],
    [4,5,6,"B"],
    [7,8,9,"C"],
    ["*",0,"#","D"]]                          # D is ENTER key

ROWS = [14,15,12,23]                          # Row pins
COLS = [24,25,8,7]                            # Column pins

cols=[0]*4
rows=[0]*4
#
# Configure columns
#
for i in range(4):                            # Conf columns
   cols[i] = OutputDevice(COLS[i])
   cols[i].on()

#
# Configure rows
#
for j in range(4):                            # Conf rows
   rows[j] = InputDevice(ROWS[j])

#
# This function reads a key from the keypad
#
def Get_Key():
   while True:
      for j in range(4):
         cols[j].off()                               # Set col j to 0
         for i in range(4):                          # For all rows
             if rows[i].value == 0:                  # Is row 0?
```

```
                    return (KEYPAD[i][j])             # Return key
                    while rows[i].value == 0:
                       pass
          cols[j].on()                                # col back to 1
          time.sleep(0.05)                            # Wait 0.05s

#
# =============== AD9850 functions ==================
#

#
# AD9850 module connections
#
FQ_UD = 26                                    # FQ_UD pin
W_CLK = 19                                    # W_CLK pin
RESET = 13                                    # RESET pin
DATA = 6                                      # DATA pin
              # GPIO26 is CS output

#
# Configure AD9850 pins as outputs
#
fq_ud = OutputDevice(FQ_UD)
w_clk = OutputDevice(W_CLK)
reset = OutputDevice(RESET)
data = OutputDevice(DATA)

#
# Set all outputs to 0 at the beginning
#
fq_ud.off()                                   # FQ_UD = 0
w_clk.off()                                   # W_CLK = 0
reset.off()                                   # RESET = 0
data.off()                                    # DATA = 0

#
# This function sends a pulse to the pin specified in the argument
#
def SendPulse(GPIOpin):
   GPIOpin.on()                               # Send 1
   GPIOpin.off()                              # Send 0
   return

#
# This function sends bits of a byte of data to the AD9850 module
#
```

```
def SendByte(DataByte):
    for k in range(0, 8):                              # Do 8 times
        p = DataByte & 0x01                            # Get bit 0
        if p == 0:
            data.off()
        else:
            data.on()
        SendPulse(w_clk)                               # Send clock
        DataByte = DataByte >> 1                       # Get next bit
    return

#
# This function puts AD9850 module into serial mode
#
def SetSerialMode():
    SendPulse(reset)                                   # Pulse RESET
    SendPulse(w_clk)                                   # Pulse W_CLK
    SendPulse(fq_ud)                                   # Pulse FQ_UD
    return

#
# This function loads the 40-bit data to the AD9850. 32-bit
# frequency data is sent, then the power down bit is sent as 0,
# two Control bits are sent as 0, and then the 5 phase bits are
# sent as 0
#
def LoadFrequency(frequency):
    f = int(frequency * 4294967296 / 125000000) # See book
    for p in range(0, 4):                              # Do 4 times
        SendByte(f & 0xFF)                             # Send Low byte
        f = f >> 8                                     # Get next byte
    SendByte(0x00)                                     # Send remaining bits
    SendPulse(fq_ud)                                   # Terminate serial
    return

Total = 0
mylcd.clear()                                          # Clear LCD
mylcd.move_to(0, 0)                                    # To (0,0)
mylcd.putstr("Frequency (Hz):")                        # Heading
mylcd.move_to(0, 1)                                    # To (0,1)

try:

  while True:
    mylcd.move_to(0, 1)
    flag = 0
```

```
    while flag == 0:
        key = Get_Key()                         # Get a key
        if key != "D":                          # Is it ENTER?
          if str(key).isnumeric():              # IS it numeric?
             mylcd.putstr(str(key))             # Display key
             N = int(key)
             Total = 10*Total + N               # Total so far
        else:                                   # ENTER detected
          frequency = Total                     # Get freuency
          Total = 0
          flag = 1
          mylcd.putstr("  - OK")                # Display OK
        time.sleep(0.2)                         # WAit 0.2 sec

    SetSerialMode()                             # Select serial mode
    LoadFrequency(frequency)                    # Load register
    time.sleep(5)
    mylcd.move_to(0, 1)
    mylcd.putstr("                ")

except KeyboardInterrupt:                       # Keyboard Cntrl+C
    SendPulse(reset)                            # Stop AD9850
```

Figure 4.43: Program listing.

Figure 4.44 shows the LCD before entering the frequency, after entering frequency and after pressing the **D** (ENTER) key respectively.

Figure 4.44: Before entering the frequency.

The project was constructed on a breadboard as shown in Figure 4.45 and connections were made to the Raspberry Pi 5 and other components using jumper wires. Interested readers can design a PCB for the project.

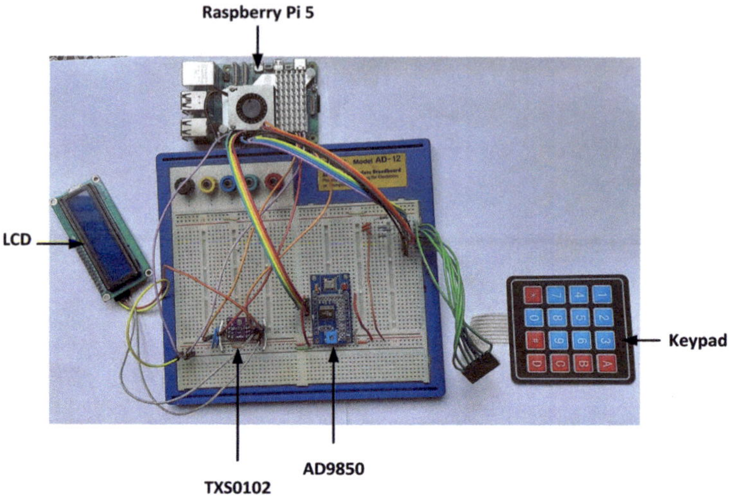

Figure 4.45: Project constructed on a breadboard.

Figure 4.46 shows the generated sine waveform at pin **Sine-wave pin 1** of the AD9850. It is clear that the frequency is exactly 2000 Hz. Figure 4.47 shows the square wave output at pin **Square-wave pin 1**.

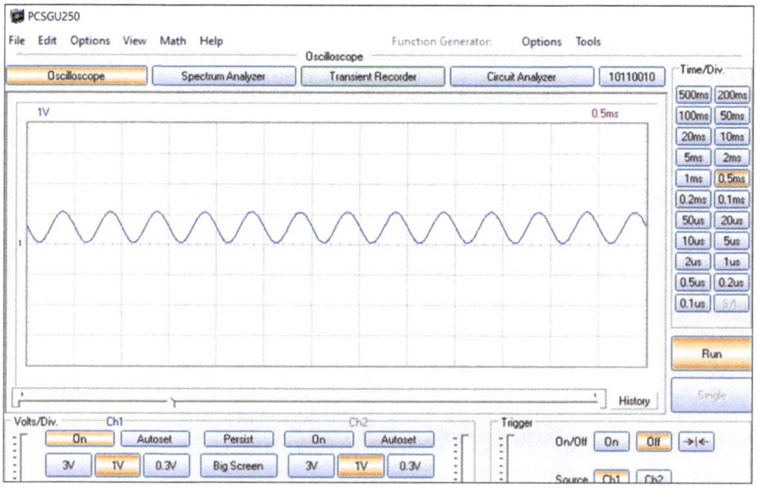

Figure 4.46: Sine wave output.

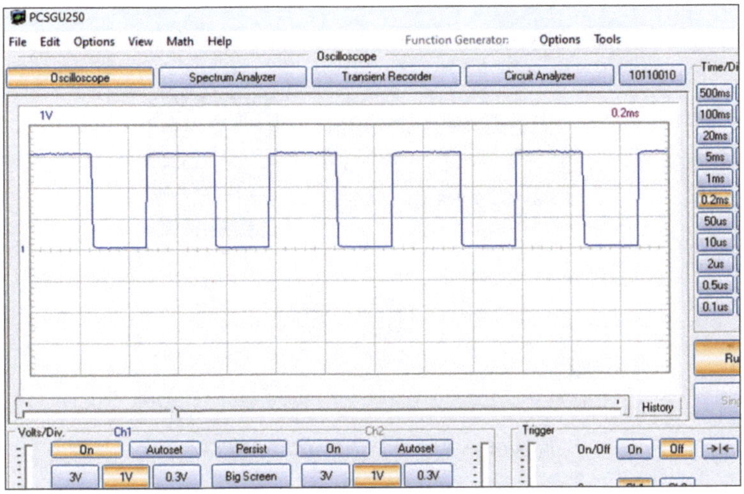

Figure 4.47: Square wave output.

4.8.2 Starting the program automatically at boot time

You may want to start the program automatically when you power up your Raspberry Pi 5, or after a reboot. The steps are as follows:

- Use the nano editor to edit file /etc/rc.local

  pi@raspberrypi:~ $ **sudo nano /etc/rc.local**

- Insert your program name to the end of the file, terminating the filename with the **&** character so that the program runs at the background when Raspberry Pi 5 starts:

  /usr/bin/python /home/pi/keypadsig.py &

Remember to remove the program from /etc/rc.local if you decide not to start the program automatically.

4.8.3 Boxing the project

You may like to box the project so that it can be used in your station as a sine wave/square wave signal generator. Figure 4.48 shows the front view of a possible box.

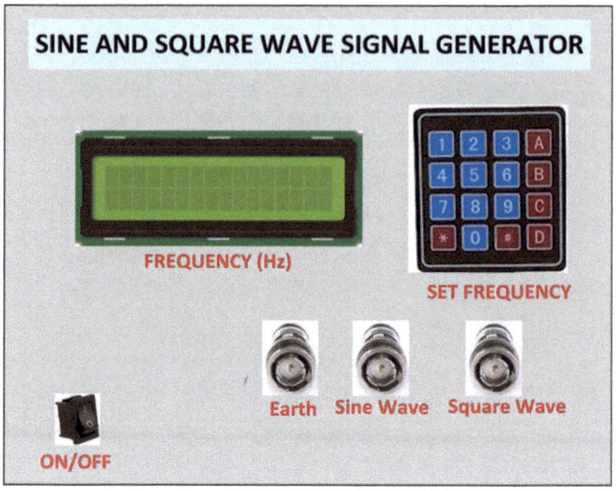

Figure 4.48: Boxing the project.

4.9 Project 8: Morse Code Exerciser with Rotary Encoder and LCD to Set wpm Value

Description: In this project we will design a Morse code exerciser where the Morse code of random characters and numbers are generated and output to a buzzer. The timing of the code is set using a rotary encoder together with an I²C LCD.

Morse code

The information given here may not be new to some readers, but it is given for completeness.

The timing in Morse code is as follows:

- A *dit*: 1 unit
- A *dah*: 3 units
- Character spacing between *dit* and *dah* of a character: 1 unit
- Character spacing between the characters of a word: 3 units
- Word spacing: 7 units

The speed of a Morse code message conveyed over radio or telegraph is specified by how many words per minute the operator can send or receive. In most amateur radio exams, candidates are expected to send and receive at least 12 words per minute (wpm).

The word **PARIS** is used as the standard word, which consists of 50 units of time:

P:	.--.	1 1 3 1 3 1 1 (3)	14 units
A:	.-	1 1 3 (3)	8 units
R:	.-.	1 1 3 1 1 (3)	10 units
I:	..	1 1 1 (3)	6 units
S:	...	1 1 1 1 1 [7]	12 units

Where () is the inter-character time, and [] is the inter-word time. Notice that at the end of a character, you do not insert a dot time, but instead insert an inter-character time (3 units). Similarly, at the end of a word, not dot time is inserted but an inter-word time (7 units).

Total units = 50

The formula for the words per minute can be calculated as follows:

Let **wpm** = words per minute
Then seconds per word is given by: **spw** = 60 / wpm

If we take *dits* per word to be 50 as the standard, then

Seconds per *dit*, **spd** = 60 / (50wpm), or spd = 1.2/wpm

We can specify this in milliseconds, giving:

Milliseconds per dit, **mpd** = 60000 / (50 wpm) = 1200/wpm

Therefore, given the required words per minute, we can calculate the time of each dit in milliseconds as:

mpd = 1200/wpm

For example, if the required words per minute is 12, then

mpd = 1200/12 = 100 milliseconds

Meaning we have to allow 100 ms for the basic unit of timing. Similarly, if the required words per minute is 20 then the bit timing should be 1200/20 = 60 ms.

Block Diagram: Figure 4.49 shows the block diagram of the project. There are two types of buzzers available: passive, and active. Passive buzzers can give sounds at different frequencies and input to such buzzers are usually signals at the required frequencies. An active buzzer on the other hand generates a single tone of sound when a logic 1 is applied across its terminals. In this project, an active buzzer is used for simplicity.

Figure 4.49: Block diagram of the project.

4.9.1 Rotary encoder

A rotary encoder is a device that looks like a potentiometer. It senses the rotation and direction of the spindle and the attached knob. The device has two internal contacts that make and break a circuit as the knob is turned. As the knob is turned, a click is felt so the user knows the device has been rotated by one position. With simple logic, you can determine the direction of rotation.

A rotary encoder has the following pins:

GND: power supply ground.
Vcc (+): power supply.
CLK: This is an output pin used to determine the amount of rotation. Each time the knob is rotated by one click in either direction, the CLK output goes to HIGH and then LOW.
DT: This is an output similar to CLK pin, but it lags the CLK by 90 degrees. This output is used to determine the direction of rotation.
SW: This is an active-Low pushbutton. When the knob is pushed, the voltage goes logic Low.

In this project, each rotation (i.e., click) of the knob will increment (or decrement) the words per minute count by 1. Turning the knob in one direction will increment the count by one, while turning it in the other direction will decrement it by one. When the required value is reached, the user has to push the knob so that the program starts generating random characters of Morse code on the buzzer.

Circuit Diagram: Figure 4.50 shows the circuit diagram of the project. The I^2C LCD is connected to Raspberry Pi 5 as in the previous LCD based projects, where the SDA and SCL pins are connected to GPIO2 and GPIO3 respectively. The buzzer is connected to pin GPIO26 through a transistor switch. The rotary encoder is connected as follows:

Rotary encoder	GPIO pin
CLK	GPIO19
DT	GPIO13
SW	GPIO6
GND	GND
Vcc(+)	3.3V

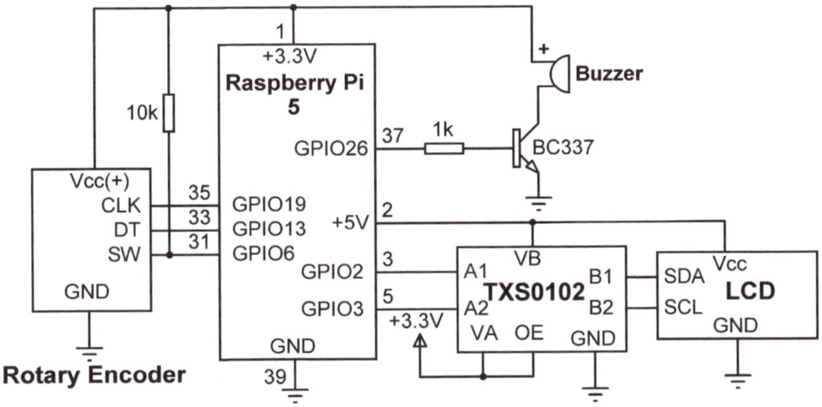

Figure 4.50: Circuit diagram of the project.

Program Listing: Figure 4.51 shows the program listing (Program: **Morse.py**). At the beginning of the program the LCD library is imported and the type of LCD is defined. Then the interface between the rotary encoder and the Raspberry Pi 5 are defined and pins CLK, DT, and SW of the rotary encoder are configured as inputs.

The words per minute value increments or decrements as the user turns the knob of the rotary encoder clockwise or anticlockwise respectively, where the minimum value is 1. The LCD displays the WPM and expects the user to select the required WPM by rotating the encoder knob. Pushing in the rotary encoder knob starts the program loop where random characters are generated and their Morse codes are sent to the buzzer. The LCD displays the characters in real time as they are generated randomly by the program.

You can start the program from the command line as:

pi@raspberrypi:~ $ **python Morse.py**

```
#-------------------------------------------------------------------
#                    MORSE CODE EXERCISER
#                    ====================
#
# This is a Morse code exercise program. The program generates
# random characters which are converted into Morse code and sent
# to the buzzer. In this program the speed is set using a rotary
# encoder and an I2C LCD. Turning the rotary encoder know increments
# or decrements the wpm. Pushing in the knob start generating the
# Morse codes. The generated characters are displayed at the bottom
# row of the LCD
#
# Author: Dogan Ibrahim
# File  : Morse.py
# Date  : February, 2024
```

```
#-----------------------------------------------------------------------
from gpiozero import InputDevice, OutputDevice
import time
import random

#
# Import LCD libraries
#
from lcd_api import LcdApi
from i2c_lcd import I2cLcd

I2C_ADDR = 0x27
I2C_NUM_ROWS = 2
I2C_NUM_COLS = 16

mylcd = I2cLcd(1,I2C_ADDR,I2C_NUM_ROWS,I2C_NUM_COLS)
mylcd.clear()                                           # clear LCD

Buzzer = 26                                             # Buzzer pin

#
# Rotary encoder connections
#
CLK = 19                                                # CLK pin
DT = 13                                                 # DT pin
SW = 6                                                  # SW pin

#
# Configure buzzer and rotary encoder
#
buzzer = OutputDevice(Buzzer)
clk = InputDevice(CLK)
dt = InputDevice(DT)
sw = InputDevice(SW)

buzzer.off()                                            # Disable Buzzer

#
# Morse code table as a dictionary
#
MorseCode = {' ': ' ',
    '0': '-----',
    '1': '.----',
    '2': '..---',
    '3': '...--',
    '4': '....-',
```

```
    '5': '.....',
    '6': '-....',
    '7': '--...',
    '8': '---..',
    '9': '----.',
    ':': '---...',
    ';': '-.-.-.',
    '?': '..--..',
    '/': '-..-.',
    '.': '.-.-.-',
    ',': '--..--',
    'A': '.-',
    'B': '-...',
    'C': '-.-.',
    'D': '-..',
    'E': '.',
    'F': '..-.',
    'G': '--.',
    'H': '....',
    'I': '..',
    'J': '.---',
    'K': '-.-',
    'L': '.-..',
    'M': '--',
    'N': '-.',
    'O': '---',
    'P': '.--.',
    'Q': '--.-',
    'R': '.-.',
    'S': '...',
    'T': '-',
    'U': '..-',
    'V': '...-',
    'W': '.--',
    'X': '-..-',
    'Y': '-.--',
    'Z': '--..'}

#
# Output a dot
#
def Dot():
    buzzer.on()
    time.sleep(DotTime)
    buzzer.off()
    time.sleep(DotTime)
```

```
#
# Output a dash
#
def Dash():
    buzzer.on()
    time.sleep(DashTime)
    buzzer.off()
    time.sleep(DashTime)

wpm = 1                                         # Default value
ClkOldState = clk.value                         # Get CLK state
flag = 1
mylcd.clear()                                   # Clear LCD
mylcd.move_to(0, 0)                             # At (0,0)
mylcd.putstr("Enter WPM:")                      # Heading
mylcd.move_to(0, 1)
mylcd.putstr("1")

try:

#
# Get the required words per minute. User turns the rotary
# encoder know to set the wpm. Also,  calculate timings
#
    while flag == 1:
      ClkState = clk.value
      DTState = dt.value
      if ClkState != ClkOldState and ClkState == 1:
         if DTState != ClkState:
            wpm = wpm + 1
         else:
            wpm = wpm - 1
            if wpm == 0:
               wpm = 1
         mylcd.move_to(0, 1)
         mylcd.putstr("      ")
         mylcd.move_to(0, 1)
         mylcd.putstr(str(wpm))
      ClkOldState = ClkState

      SWState = sw.value                        # Get state of SW
      if SWState == 0:                          # Knob pushed in?
         flag = 0                               # Clear flag

#
```

```
# We come here when the user pushes in the rotary encode knob
#
   UnitTime = 1.2/wpm
   DotTime = UnitTime
   InterCharTime = 2 * UnitTime
   DashTime = 3 * UnitTime
   WordTime = 6* UnitTime

   values = MorseCode.values()
   values_list = list(values)
   keys = MorseCode.keys()
   keys_list = list(keys)

   mylcd.clear()                                        # Clear LCD
   mylcd.move_to(0, 0)                                  # At (0, 0)
   s = "Working: " + str(wpm) + " wpm"                  # Heading
   mylcd.putstr(s)
   mylcd.move_to(0, 1)                                  # At (1, 0)
   col = 0

#
# The Morse codes are generated in the loop below
#
   while True:                                  # Do forever
      r = random.randint(1, 42)                 # Random number
      c = values_list[r]                        # Get a value
      d = keys_list[r]                          # Get its key
#      print(d, end="", flush=True)             # Display key
      mylcd.putstr(d)                           # Display chars
      col = col + 1
      if col > 16:
         col = 0
         mylcd.move_to(0, 1)
         mylcd.putstr("                )
         mylcd.move_to(0,1)
      for code in c:                            # Do for code
         if code == '-':                        # If dash
            Dash()
         elif code == '.':                      # if dot
            Dot()
      time.sleep(InterCharTime)

except KeyboardInterrupt:
   print("")
```

Figure 4.51: Program listing.

Figure 4.52 shows example displays on the LCD. Figure 4.53 shows the project built on a breadboard.

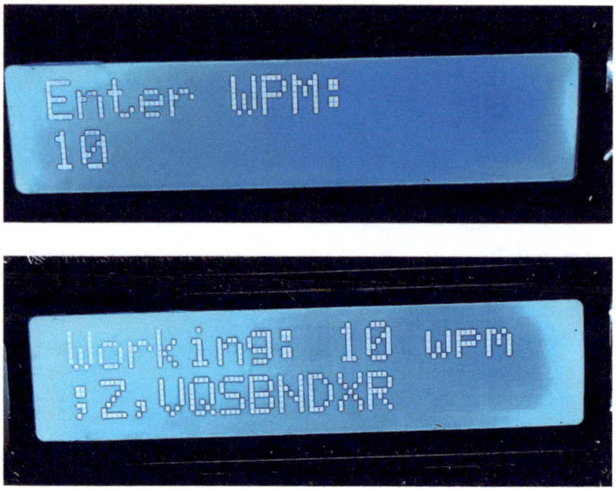

Figure 4.52: Example displays.

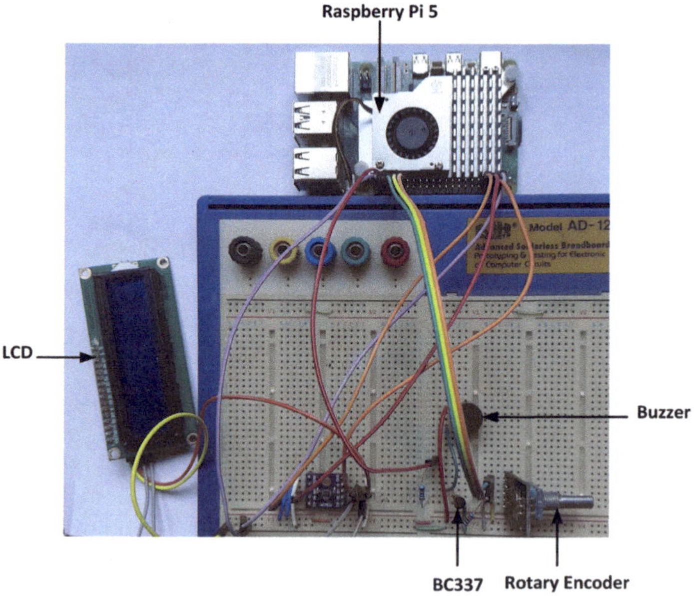

Figure 4.53: Project built on a breadboard.

4.9.2 Boxing the project

You may like to box the project so that it can be used in your station as a Morse code exerciser. Figure 4.54 shows the front view of a possible box.

Figure 4.54: Boxing the project.

4.10 Project 9: Morse Decoder

Morse code is still used by many amateur radio operators. Although it is much more fun to decode the code manually by listening to it, there are many amateurs who may prefer to use a computer to decode the code and have it displayed on a LCD or on a computer screen.

In previous project, you have learned how to generate Morse code using the Raspberry Pi 5. There, the project was about setting the required word-per-minute and then generating random characters and sending their Morse codes to a buzzer as well as displaying these characters on LCD.

In this project you will develop a Morse code decoder which will receive audible code through a microphone and then translate the code into text and display it on the screen in real time. The program adjusts to the transmission speed of the code automatically.

Just to review, the Morse code timing (i.e., the speed) is as follows:

- A Dit: 1 unit
- A Dah: 3 units
- Character spacing between dit and dah of a character: 1 unit
- Character spacing between the characters of a word: 3 units
- Word spacing: 7 units

The speed of a Morse code is specified by how many words per minute can be sent or received. In most Amateur radio exams candidates are expected to send and receive at least 12 words per minute.

The word PARIS is used as the standard word when calculating the speed. This word consists of 50 units of time:

P:	.--.	1 1 3 1 3 1 1 (3)	14 units
A:	.-	1 1 3 (3)	8 units
R:	.-.	1 1 3 1 1 (3)	10 units
I:	..	1 1 1 (3)	6 units
S:	...	1 1 1 1 1 [7]	12 units

Where () is the inter-character time, and [] is the inter-word time. Notice that at the end of a character, we do not insert a dot time, but we insert an inter-character time (3 units). Similarly, at the end of a word, we do not insert a dot time, but we insert an inter-word time (7 units).

Knowing the words per minute (WPM), you can calculate the Dit (or bit) time in seconds as follows:

$$\text{Dit time} = 60 / (50 \times \text{wpm})$$

Thus, for example, at 20 wpm, the basic Dit time is given by: 60 / (50 × 20) = 60 ms.

In this project, Morse code is received through a microphone as audio and the program initially estimates the Dit timing. The other timings are then automatically calculated and the received codes are translated to text and displayed on LCD.

Block diagram: Figure 4.55 shows the block diagram of the project. An audio amplifier with on-board microphone module (**Mic click board,** manufactured by Mikroelektronika) is used to receive the audible Morse code. The audio code is then amplified and sent to a tone decoder IC (LM567), which generates binary High and Low values based on the input tone. The output of the tone generator is connected to a GPIO pin of the Raspberry Pi 5. The translated code is then displayed.

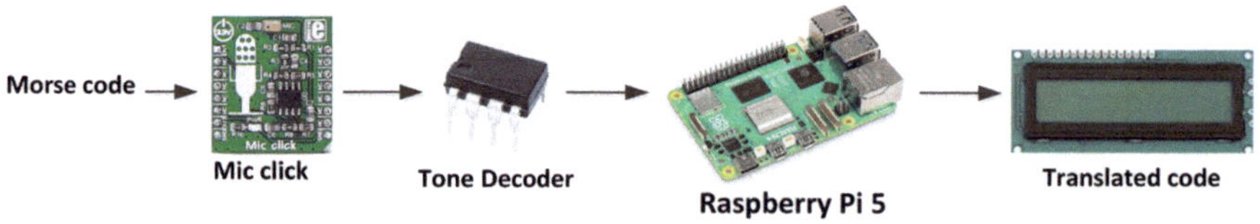

Figure 4.55: Block diagram of the project.

The Mic click board

The **Mic click** is an audio amplifier board (see www.mikroe.com) that carries a SPQ0410HR5H-B surface mount silicon microphone with maximum RF protection. The board (Figure 4.56) is designed to run on a 3.3 V power supply. Using the SiSonic™ MEMS technology, the SPQ0410HR5H-B consists of an acoustic sensor, a low noise input buffer, and an analog output amplifier. **MaxRF protection** prevents RF noise in traces from getting into the mic output.

Figure 4.56: The Mic click board.

The basic features of the Mic click board are:

- Supply current: 120 μA (typical)
- Sensitivity: -42 dBV/Pa
- Signal to noise ratio: 59 dB(A)
- Total harmonic distortion: 1%
- Output impedance: 400 ohms
- Analog output
- Direction: Omnidirectional

Pin 1 is the output voltage. +3.3 V power supply must be connected to pin 7 of the board.

The tone decoder

Figure 4.57 shows pin layout of the LM567 tone decoder IC. **LM567** is a high-stability low-frequency integrated phase-locked loop decoder. Due to its good noise suppression ability and center frequency stability, it is widely used in the decoding of various communication equipment and the demodulation circuit of AM and FM signals. The chip is also used in circuits such as touch tone decoding, ultrasonic controls, precision oscillator, frequency monitoring and control, paging detectors etc. The LM567 tone decoder is a device capable of detecting whether an input signal is within a selectable detection range. The device has an open collector transistor output, so an external resistor is required to reach the appropriate logic levels. When the input signal is in the detection band, the device output changes to the LOW state. The internal free operating frequency of the VCO defines the center frequency of the detection band. An external RC filter is required to adjust this frequency. The bandwidth in which the device will detect the desired frequency depends on the capacity of the loop filter terminal. Usually, a 1μF capacitor is connected to this pin. Pin descriptions of the LM567 are shown in Table 4.1.

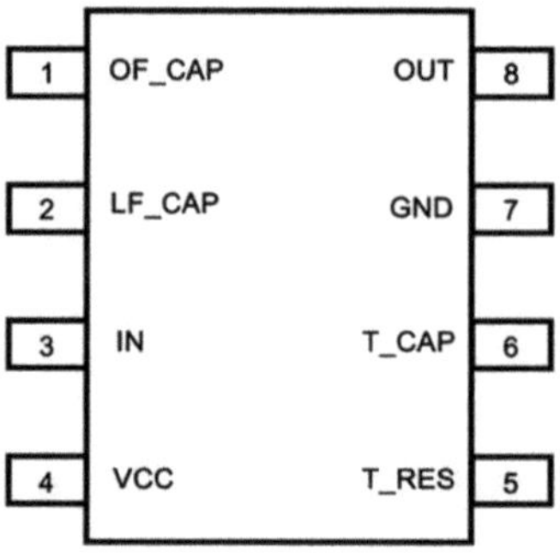

Figure 4.57: Pin layout of LM567.

PIN		TYPE	DESCRIPTION
NAME	NO.		
GND	7	P	Circuit ground.
IN	3	I	Device input.
LF_CAP	2	I	Loop filter capacitor pin (LPF of the PLL).
OUT	8	O	Device output.
OF_CAP	1	I	Output filter capacitor pin.
T_CAP	5	I	Timing capacitor connection pin.
T_RES	6	I	Timing resistor connection pin.
VCC	4	P	Voltage supply pin.

Table 4.1: LM567 pin descriptions.

The LM567 IC has the following basic features:

- Operating voltage: 4.5 V to 9.0 V
- Quiescent power supply current: 7 mA (typical)
- Active power supply current: 12 mA (typical)
- Smallest detectable input voltage: 20 mv_{rms}
- Highest center frequency: 500 kHz
- Center frequency stability: 35 ± 60 ppm/ºC
- Output saturation voltage: 0.2 V (typical)
- Switching speed: center frequency / 20

Circuit diagram

Figure 4.58 shows the circuit diagram of the project. The Morse code audio signal is applied to pin 3 through a 470 nF capacitor. The Output is open-collector and taken from pin 7.

The center frequency of the LM567 tone decoder is equal to the free-running frequency of the voltage-controlled oscillator. In order to set this frequency, external components should be placed externally. The component values are given by:

$$f_0 = \frac{1.1}{R_1 C_1}$$

Where, $R1$ = Timing Resistor at pin 5, $C1$ = timing capacitor at pin 6. Using a 10-kΩ resistor in series with a 10-kΩ potentiometer and 0.1-µF capacitor, the center frequency can take the following values as the potentiometer is varied:

at 10 kΩ $f_0 = 550\ Hz$

at 5 kΩ $f_0 = 733\ Hz$

at 1 kΩ $f_0 = 1000\ Hz$

at 0 kΩ $f_0 = 1100\ Hz$

The frequency range f_0 = 500 Hz to about f_0 = 800 Hz should be a good choice with the potentiometer arm between full and half turn.

To eliminate undesired signals that could trigger the output stage, a post detection filter is featured in the LM567C. This filter consists of an internal 4.7 kΩ resistor and an external capacitor. Although typically external capacitor value is not critical, it is recommended to be at least twice the value of the loop filter capacitor. If the output filter capacitor value is too large, the turn-on and turn off-time of the output will present a delay until the voltage across this capacitor reaches the threshold level.

The bandwidth depends on the capacitor connected to pin 2 (C_2) and is given as a percentage of the center frequency, where (assuming Vi <= 200 mV_{rms}):

$$BW = 1070\sqrt{\frac{Vi}{f_0 C_2}}\ \%\ of\ f_0$$

Where Vi is in volts rms, f_0 is Hz, and C_2 in µF.

For example, with f_0 = 600 Hz, and C_2 = 1 µF, the BW will be about 20% of the center frequency which will be approximately 120 Hz. A smaller C_2 value will increase the bandwidth and a larger value will decrease it.

The output filter is determined by capacitor C_3. Although it is recommended to choose a value to be at least twice the value of C_2, it is found that about 1 µF gives good results.

The switching speed of LM567 depends on the center frequency. With f_0 = 550 Hz, the switching speed is 550/20 = 27.5 and with f_0 = 800 Hz, the switching speed is 800 / 20 = 40. It is therefore necessary to choose high center frequencies for faster switching rates. Because the output of LM567 is open-collector, it is necessary to connect a pull-up resistor to this pin as shown in Figure 6.169. An LED is connected in series with the pull-up resistor and this LED flashes to indicate the received Morse code output by the tone decoder.

Figure 4.59 shows the frequency detection where the output changes state (goes from High to Low) when the input reaches the center frequency.

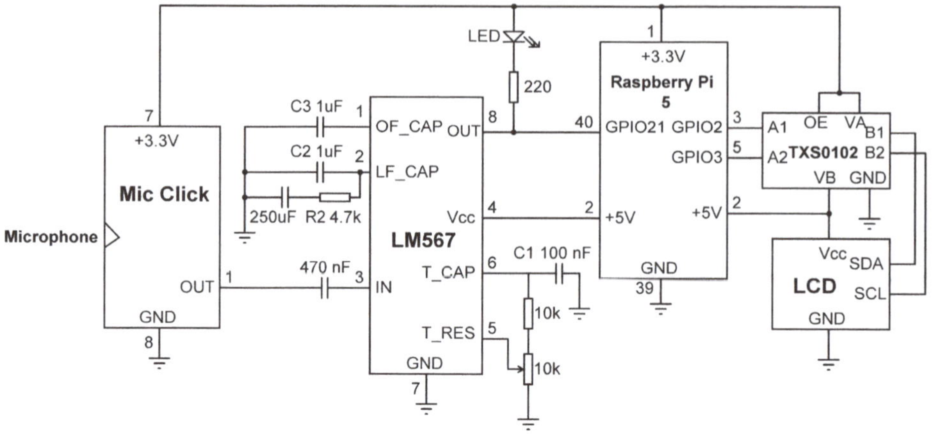

Figure 4.58: Circuit diagram of the project.

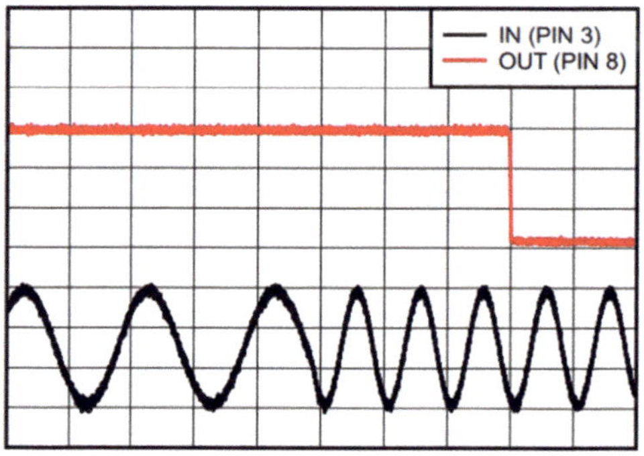

Figure 4.59: Frequency detection.

The program listing is shown in Figure 4.60 (Program: **MorseDecoder.py**). In this program, the Morse code is embedded into the binary version of numbers from 2 to 63 as described by Budd Churchward (WB7FHC, VK2IDL_Morse_Decoder_2.7, see this link: https://github.com/ideal54/VK2IDL_CW_Decoder). The operation is as follows:

Initially 1 is stored in a variable, say **MyNum**. After a dit or dah is received, this number is the shifted left. If a dit is received then **MyNum** is incremented by one, if a dah is received, just 0 is added to the number and the next dit or dah is processed. The following string is used to transform the number in **MyNum** into letters and numbers:

Morse = "##TEMNAIOGKDWRUS##QZYCXBJP#L+F VH09#8###7#####/-61#######2###3#45";

As an example, some characters and the number stored in **MyNum** are given below (remember that the leftmost 1 is always added to the number):

Morse code:	dah	dit	dahdah	dahdit	ditdah	ditdit
Letter:	T	E	M	N	A	I
MyNum (binary):	10	11	100	101	110	111
MyNum (decimal):	2	3	4	5	6	7

If we now use **MyNum** as an index into string **Morse** we will get the correct characters. e.g., Morse[2] = T, Morse[3] = E, Morse[4] = M and so on. Notice that the index positions of the **#** characters are not used. i.e., Morse[0] and Morse[1] are not used.

At the beginning of the program the libraries used are imported to the program. Text **MORSE** is displayed at the top row of the LCD for 2 seconds. The program consists of 2 functions: **Setup()** and **Decode().**

Setup(): This function receives Morse code of at least 10 dits and dahs and then finds the maximum dit time in milliseconds and stores it in variable **ditmax**. It is therefore necessary to train the program at the beginning by sending several characters (e.g., sending ABCDEF). This value is used in function **Decode()** to determine if a *dit* or a *dah* is received. The edges of the input signal are detected using **while** statements. For example, the following statements wait until the signal changes from 1 to 0. Built-in function **time.monotonic()** is used at the beginning and end of an input change in order to determine the durations in milliseconds of the input high or low times.

```
while signal.value() == 1:
       pass
```

Decode(): This is the main function of the program where the Morse code is received and transformed into letters and numbers. Variable **MyNum** is initially set to 1. After receiving a Low and a High signal, variable **MyNum** is shifted left by one position so that it contains binary number 10. The program then determines whether the Low signal was a dit or a dah. Variables **Mark** and **Space** refer to signal Low and High states respectively, where Low is when a tone is received and High is when there is no tone.

```
#------------------------------------------------------------
#                          MORSE CODE DECODER
#                          ------------------
#
# This is a Morse code decoder project implemented on the Raspberry
# Pi 5. A Mic click board, equipped with a microphone is used to
# receive the audible Morse code. This code is then fed to a LM567
# type tone decoder chip. The tone decoder trigger frequency is set
# by using a potentiometer. An LED is connected to the output of the
# tone decoder which makes it easy to adjust the frequency. It is
```

```
# recommended to use around 800 Hz audio. The Raspberry Pi 5 decodes
# the Morse code and displays in text form on an LCD
#
# Author: Dogan Ibrahim
# File  : MorseDecoder.py
# Date  : February, 2024
#-------------------------------------------------------------------
import time
from gpiozero import InputDevice
#
# Import LCD libraries
#
from lcd_api import LcdApi
from i2c_lcd import I2cLcd

I2C_ADDR = 0x27
I2C_NUM_ROWS = 2
I2C_NUM_COLS = 16

mylcd = I2cLcd(1,I2C_ADDR,I2C_NUM_ROWS,I2C_NUM_COLS)

Morse = "##TEMNAIOGKDWRUS##QZYCXBJP#L+FVH09#8###7#####/-61#######2###3#45"
mylcd.clear()
mylcd.putstr("MORSE")
time.sleep(2)
mylcd.clear()
tim = 1000

global ditmax, MyNum
maxdit = [0]*10
MyNum = 1
SIGNAL = 21                                        # At GPIO21
#
# Signal connections. Morse is input to GP21 through tone decoder
#
signal = InputDevice(SIGNAL)

#
# Find max dit and dah times by listening to some code. ditmax is
# the maximum measured dit time which is used later in the code
#
#
def Setup():
  global ditmax, dahmax
  for i in range(0, 10):
     while signal.value == 1:                      # Wait if 1
```

```
        pass
      time.sleep(0.01)
      TmrStrt = tim*time.monotonic()              # Elapsed time

      while signal.value == 0:                    # Wait if 0
        pass
      time.sleep(0.01)
      TmrEnd = tim*time.monotonic() - 20          # Elapsed time
      maxdit[i] = TmrEnd - TmrStrt
  dahmax = max(maxdit)
  ditmax = dahmax / 3

#
# This function runs continuously, decoding and displaying the
# Morse code on the LCD
#
def Decode():
  global MyNum, ditmax
  flag = 0
  if signal.value == 1:
     mylcd.putstr(" ")
  while signal.value == 1:
     pass
  time.sleep(0.01)
  while True:
     TmrStrt = tim*time.monotonic()
     while signal.value == 0:
        pass
     time.sleep(0.01)
     TmrEnd = tim*time.monotonic()
     MarkTime = TmrEnd - TmrStrt

     TmrStrt = tim*time.monotonic()
     while signal.value == 1:
        if (tim*time.monotonic() - TmrStrt) > 5 * ditmax:
           flag = 1
           break
     time.sleep(0.01)
     TmrEnd = tim*time.monotonic()
     SpaceTime = TmrEnd - TmrStrt

     MyNum = MyNum << 1
     if MarkTime < 2*ditmax:
        MyNum = MyNum + 1
     if SpaceTime > 2*ditmax:
        mylcd.putstr(Morse[MyNum])               # Display text
```

```
            MyNum = 1
        if flag == 1:
            break

Setup()

try:
   while True:
        Decode()
except:
   print("End of program")
   time.sleep(2)
```

Figure 4.60: Program listing.

Testing the program

The steps to test the program are:

- Construct the hardware.
- Apply power to the Raspberry Pi 5. Make sure that the green LED on the Mic click board is turned ON. Run the program (click the green button if using Thonny, otherwise enter **python MorseDecoder.py** at the console). Make sure that the text **MORSE** is displayed on the LCD for 2 seconds. If these do not happen then you should check your hardware and software carefully before continuing.
- Use a Morse code audio generator apps, e.g., *on your smartphone*. For example, the **Morse Code Translator** program, available on Internet (link: **https://morsecode.world/international/translator.html**) can be very helpful during testing (see Figure 4.61) using your Android smartphone.

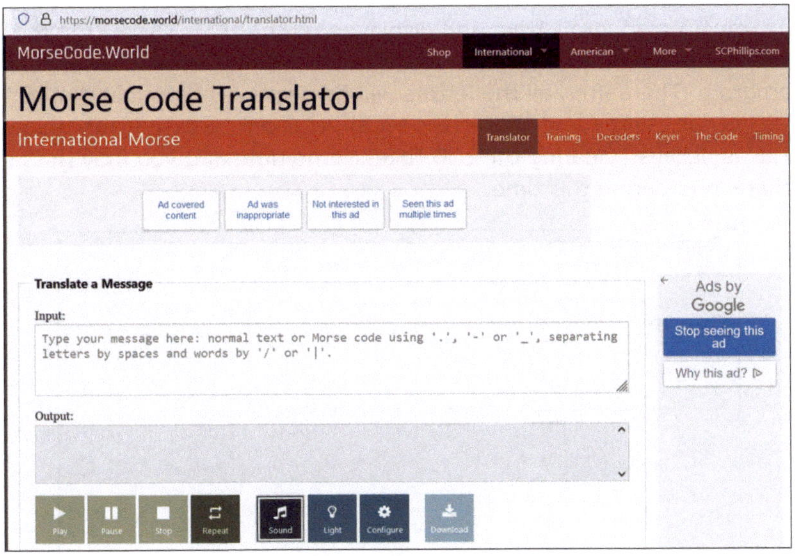

Figure 4.61: Morse code translator apps.

- Click **Configure** and set it to 12 wpm, 800 Hz tone, and maximum volume of 100 as shown in Figure 4.62.

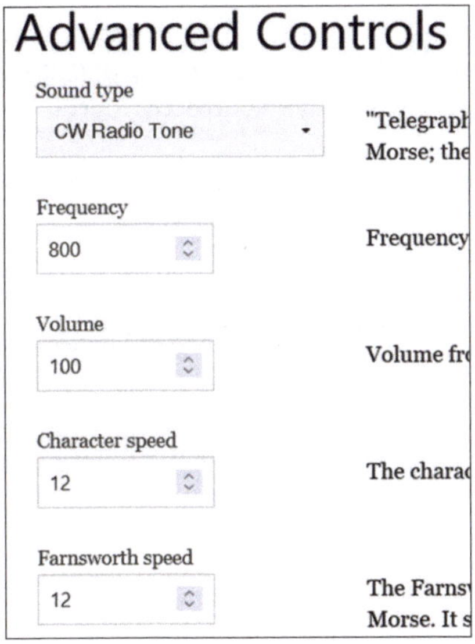

Figure 4.62: Configure the apps.

- Place your smartphone very close to the Mic click board microphone and type in some letters in window labelled **Translate a Message**. e.g., ABCDEFG (see Figure 4.63), and press **Play**. ***Turn the frequency potentiometer slowly until the LED starts flashing***. The LED will flash as the Morse code is output from your smartphone speaker. After a while, the Morse code will be transformed into letters and displayed on the LCD. Notice that the first few letters will not be displayed as they will be used to set up the speed in the program. Thereafter, all the letters will be displayed correctly. Notice that the LCD is not cleared and characters on the LCD are replaced with the new ones. This is because clearing the LCD takes some time and you may miss some characters during this time.

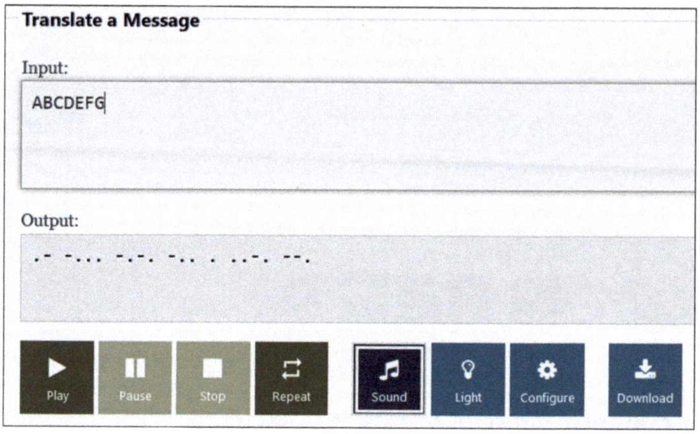

Figure 4.63: Enter some letters for testing.

Note: The program was tested at speeds up to 80 wpm. It was found that up to about 60 wpm the characters are received, transformed, and displayed correctly. Above 60 wpm most of the characters are displayed correctly, but occasional errors could occur.

It is recommended to keep the tone volume high and the audio output very close to the Mic click board microphone.

Figure 4.64 shows the project built on a breadboard. You may want to save the program so that it starts automatically after reboot (see Section 4.8.2).

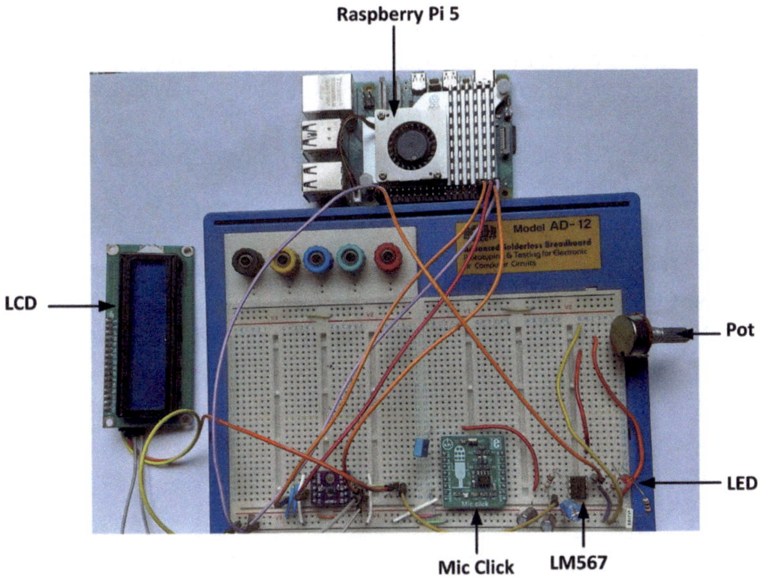

Figure 4.64: Project built on a breadboard.

4.11 Project 10: Frequency Counter

Description: There are many cases where you want to know the frequency of a signal. Frequency counters are one of the important instruments used by all amateurs. Without a frequency counter the only other option to find out the frequency of a signal is to use an oscilloscope. In this project, you will be using a Raspberry Pi 5 to design a frequency counter circuit. The frequency of the measured signal will be displayed on an I²C LCD. The signal whose frequency is to be measured is initially converted into square wave so that it is easier to interface to a processor. There are basically three methods to measure the frequency of a signal as described below.

Measuring the period: In this method, the signal is applied to one of the digital pins of the processor. The processor measures the period of the signal by using an internal timer. Here, the timer is started on the rising edge of the signal, and is stopped on the next rising edge. Knowing the period, we then take its inverse to calculate the frequency. Although this method can give accurate results, it requires a very accurate timer. Raspberry Pi 5 is not an MCU as it has an operating system and its internal timer is not suited to measuring the frequency using this method.

Measuring the pulses (internal timer): In this method, an internal timer is used as a gate. The processor starts a timer with a known duration (e.g., one second) and counts the number of pulses received during this time period. The frequency is then calculated from the knowledge of this count. For example, if the gate time is one second, then the number of counts is directly proportional to the frequency of the signal. This method required an internal processor counter, which can be updated by an external signal, and is not suited to Raspberry Pi 5.

Measuring the pulses (external timer): Here, an external counter chip is used to count the number of pulses in a given time duration. This count is then read by the processor and the frequency is easily calculated. In this project we will be using this third method.

Block Diagram: Figure 4.65 shows the block diagram of the project.

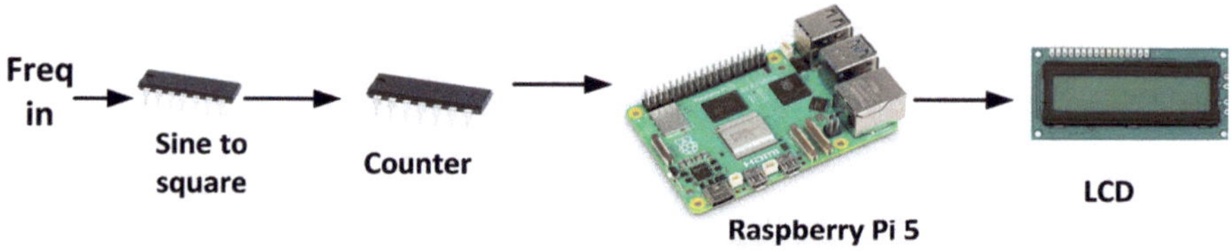

Figure 4.65: Block diagram of the project.

Circuit Diagram: In this project a type SN74LV8154 counter chip is used. The basic features of this chip are:

- Dual 16-bit counter
- Can be chained as a 32-bit counter
- 2 V to 5.5 V operating voltage
- Clear input

The nice thing about this counter is that it can be chained as a 32-bit counter for more accuracy. Additionally, it operates at 3.3 V, making it compatible with the Raspberry Pi 5 GPIO.

Figure 4.66 shows the pin configuration of the chip.

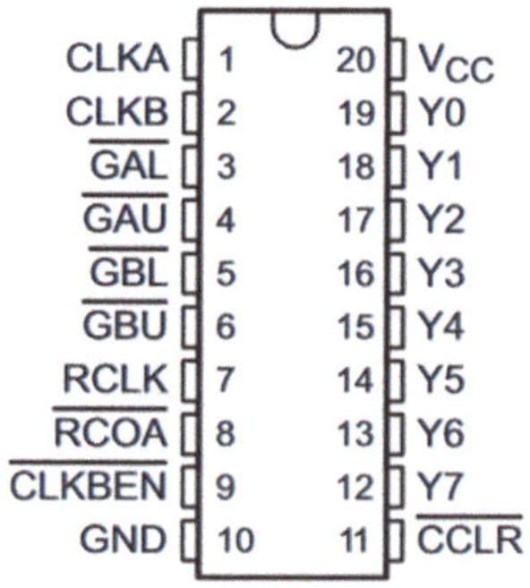

Figure 4.66: Pin configuration of SN74LV8154.

The pin descriptions are (A and B are the two identical halves of the counter):

CCLR: clear input
CLKA, CLKB: rising edge clock inputs
CLKBEN: clock B enable (active LOW)
GAL, GAU: gate A lower and upper bytes (active LOW puts A data on Y bus)

GBL, GBU: gate B lower and upper bytes (active LOW puts B data on Y bus)
GND: power supply ground
RCLK: register clock (rising edge stores data internally)
RCOA: active LOW when counter A is full, and ready to overflow to B
Vcc: power supply
Y0-Y7: Data outputs (Y0 is LSB)

The circuit diagram of the project is shown in Figure 4.67. The circuit requires a 1-Hz timing pulse for its RCLK input. On the rising edge of RCLK, the value of the 32-bit counter is stored internally so that it can be read through its Y0–Y7 outputs in 4 stages. In this project, the 1 Hz timing pulses are obtained using a GPS receiver (the *GPS click* from www.mikroe.com). Pin TP of GPS click receiver is connected to pin RCLK of SN74LV8154. It is also possible to use a 32768 Hz crystal with a 15-stage binary counter circuit as shown in Figure 4.68. This circuit is made up of a 14-bit counter and a D-type flip-flop, driven by a 32768 HZ square waveform.

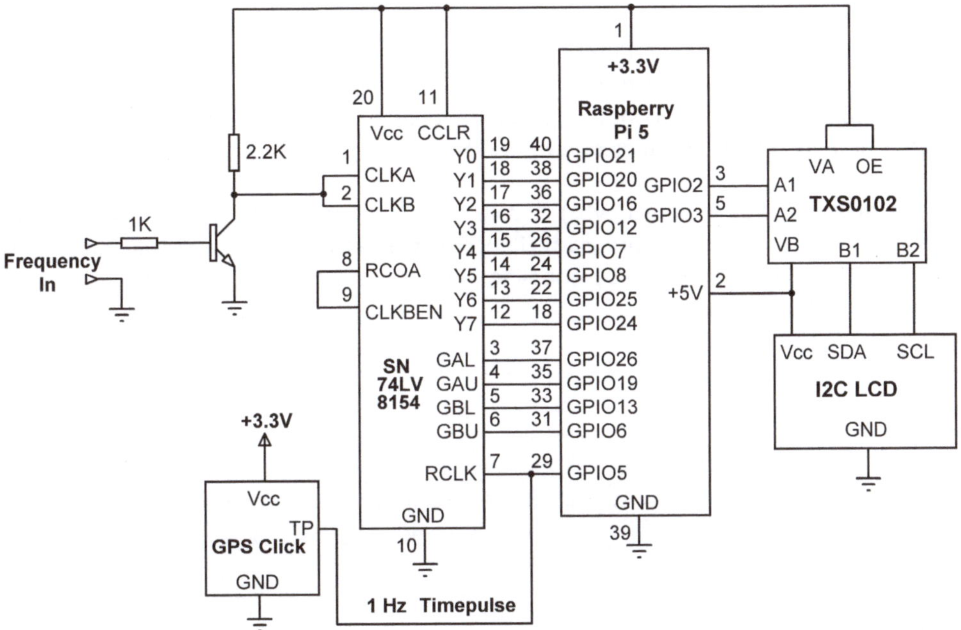

Figure 4.67: Circuit diagram of the project.

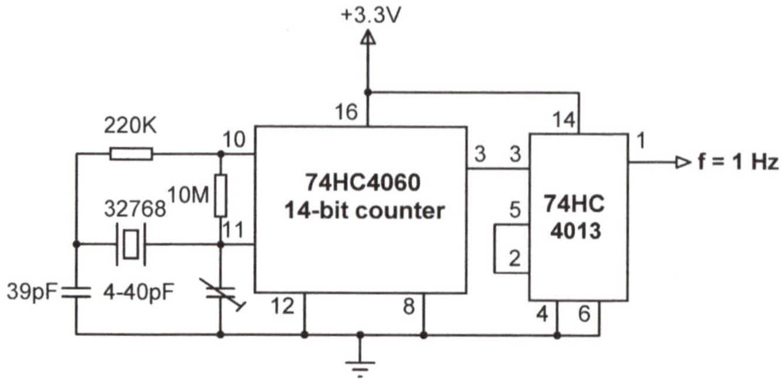

Figure 4.68: 15 stage binary counter with 32768 Hz crystal.

Program Listing: Figure 4.69 shows the program listing (Program: **freq.py**). The operation of the project is as follows: RCLK input is pulsed every second by the time pulses received from the GPS receiver (these time pulses have about 1 ms ON time). As soon the rising edge of this pulse is detected, the 32-bit value is read and stored in variable **Reading1**. The next reading is obtained on the arrival of the next pulse and this is stored in variable **Reading2**. The time difference between the two adjacent pulses is exactly one second. Therefore, the difference between the two readings (i.e., variable **difference**) is the number of counts in a second, i.e., the frequency of the waveform. By dividing this value by 1,000 you can find the frequency in kHz.

Function **GetByte** reads 8 bits (a byte) of data from the output Y0 – Y7 of the counter and returns this value to the main program. Function **GetData** reads the 32-bit data stored inside the 74LV8154 counter chip by enabling inputs GAL, GAU, GBL, and GBU. The data is returned to the main program.

```
#--------------------------------------------------------------------
#                     FREQUENCY COUNTER
#                     =================
#
# This is a frequency counter software. The ferquency is displayed
# on an LCD
#
# Author: Dogan Ibrahim
# File  : freq.py
# Date  : February, 2024
#----------------------------------------------------------------------
import time
from gpiozero import OutputDevice,InputDevice
#
# Import LCD libraries
#
from lcd_api import LcdApi
from i2c_lcd import I2cLcd
```

```
I2C_ADDR = 0x27
I2C_NUM_ROWS = 2
I2C_NUM_COLS = 16

mylcd = I2cLcd(1,I2C_ADDR,I2C_NUM_ROWS,I2C_NUM_COLS)
mylcd.putstr("Frequency Cntr")
time.sleep(2)
mylcd.clear()                                  # clear LCD

YPINS = [21, 20, 16, 12, 7, 8, 25, 24]         # Y0-Y7 pins
GPINS = [26, 19, 13, 6]                        # GA/GB pins
data = [0, 0, 0, 0]

RCLK = 5                                       # RCLK pin
disable = 1
enable = 0
pwrof32 = 2**32

gpins = [0]*4
ypins = [0]*8

for j in range(4):
   gpins[j] = OutputDevice(GPINS[j])

for j in range(8):
   ypins[j] = InputDevice(YPINS[j])            # Set as inputs

rclk = InputDevice(RCLK)

for j in range(4):                             # Disable all
   gpins[j].on()

#
# Read a byte of data from specified 8 port bits
#
def GetByte():
  total = 0
  k = 0
  for j in range(8):
     data = ypins[j].value
     data = data << k
     total = total | data
     k = k + 1
  return total

#
```

```
# Get 32-bit data from the counter
#
def GetData():
   for j in range(4):
      gpins[j].off()
      data[j] = GetByte()
      gpins[j].on()
   count = data[0] | (data[1] << 8) | (data[2] << 16) | (data[3] << 24)
   return count

while rclk.value == 1:                    # Wait if 1
   pass
while rclk.value == 0:                    # Wait if 0
   pass

Reading1 = GetData()                      # Read Reading1
while rclk.value == 1:
   pass

mylcd.clear()
mylcd.move_to(0, 0)
mylcd.putstr("Frequency (kHz):")          # Heading

try:

   while True:                            # Do forever
      while rclk.value == 0:              # While 0
         pass

      Reading2 = GetData()                # Read Reading2
      while rclk.value == 1:              # While 1
         pass

      difference = Reading2 - Reading1            # Caculate diff
      if difference < 0:
         difference = difference + pwrof32
      frequency = difference / 1000               # Frequency (kHz)
      frequency = str(frequency)[:12]             # As string
      mylcd.move_to(0, 1)
      mylcd.putstr("              ")
      mylcd.move_to(0, 1)
      mylcd.putstr(frequency)                     # Display
      Reading1 = Reading2

except KeyboardInterrupt:
    exit(0)
```

Figure 4.69: Program listing.

A resistive potential divider circuit can be used at the input for large inputs. Also, binary divider circuits can be used at the input of the circuit to increase the frequency range of the meter. For example, by using a 4-bit binary dividing circuit the useful range of the meter can be extended by a factor of 16 (you have to make sure that the components you use can work at the chosen higher frequencies).

Figure 4.70 shows an example output of the program on the LCD. In this example, the input frequency was 500 kHz.

Figure 4.70: Example output.

Testing the project
The steps are:

- Construct the hardware and connect the Raspberry Pi 5 to power supply.
- Connect an external antenna to the GPS receiver and make sure that the antenna can see the direct sky. The green LED should turn ON to indicate that power is applied to the GPS click board. Wait until the red LED on the GPS click board flashes every second. This may take several minutes.
- Use a sine wave generator and set its output frequency e.g., to 100 kHz, 3 V peak-to-peak amplitude and adjust the offset so that the waveform is all positive.
- Connect the output of the sine wave generator to the input (earth and transistor Base) of your frequency counter.
- You should see the frequency displayed in kHz on the LCD.

You can start the project automatically after reboot if you wish, as described in Section 4.8.2.

4.12 Project 11: FM Radio with Raspberry Pi 5

Description: FM radios are very popular nowadays as there are many stations to choose from on the FM band, Aso, practically all radios are equipped with the FM band. The sound quality of FM radios is also superb and most stations around the world provide stereo transmissions.

This is an FM radio project based on the TEA5767 FM radio module, is a low-cost FM radio device costing around $10. The module (Figure 4.71) is interfaced to the external world via its I²C pins (SDA and SCL). It has 4 pins, the remaining two pins are for the power supply and ground. Two small jack sockets are provided on the module; one for the antenna, and the other one for an earphone. A small solid antenna with a jack connector is provided with the module. All the user has to do is connect an earphone and program the module to the required frequency. The antenna is connected to the left hand socket when facing the component side of the module, and the earphone to the other socket.

Figure 4.71: TEA5767 FM radio module.

The TEA5767 is a single-chip electronically-tuned FM stereo radio for low-voltage applications, incorporating IF selectivity and demodulation. The module has the following features (further information can be obtained from the *TEA5767 Product Data Sheet*):

- Operation in the US/European FM band of 87.5 MHz to 108 MHz, and also Japanese FM band of 76 MHz to 91 MHz
- RF Automatic Gain Control (AGC) circuit
- Phase-locked loop (PLL) synthesizer tuning system
- Crystal reference frequency oscillator
- I²C bus interface requiring only 2 pins
- Software mute
- Signal dependent mono to stereo blend
- Signal dependent High Cut Control (HCC)
- Autonomous search tuning function
- Standby mode
- Two software programmable ports
- 2.5 V to 5 V operation
- 8.4 mA typical analog supply current, 3 mA digital supply current
- Stereo channel separation of 30 dB
- Typically 0.4% harmonic distortion

The default I²C address of the module is 0x60. Data sequence has to be in this order:

address, byte 1, byte 2, byte 3, byte 4 and byte 5. The Least Significant Bit (LSB) = 0 of the address indicates a WRITE operation to the TEA5767HN. Bit 7 of each byte is considered as the Most Significant Bit (MSB) and has to be transferred as the first bit of the byte. The data becomes valid bitwise at the appropriate falling edge of the clock. A STOP condition after any byte can shorten transmission times. At Power-on reset the mute is set, and all other bits are set to LOW.

The TEA5767 is programmed by loading 5 bytes of data to its registers via the I^2C bus. Table 5 to Table 16 of the data sheet specifies the data to be loaded for various device configurations. For example, the first data byte consists of the following bits:

Bit 7: mute bit. Setting this bit mutes the radio. It must be cleared for normal operation.

Bit 6: search mode bit. Setting this bit puts the chip into search mode. It must be cleared for normal operation.

Bit 0 – 5: higher 6 bits of the required frequency.

The second byte contains the 8 low bits of the required frequency.

The frequency is specified as 14 bits by loading the PLL register. The data to be loaded to the PLL register is given by:

$$N = \frac{4 \times (f_{RF} + f_{IF})}{f_{ref}}$$

Where

N = decimal value of PLL word.

f_{RF} is the wanted tuning frequency in Hz.

f_{IF} is the intermediate frequency in Hz (225 kHz).

f_{ref} is the reference frequency in Hz (32768 for the 32.768 kHz crystal).

As an example, to receive the broadcast at 100 MHZ, the required PLL word is:

$$N = \frac{4 \times (100 \times 10^6 + 225000)}{32768} = 12234$$

This data must be divided into two bytes and the upper 8 bits loaded as data byte 1 (with mute and search cleared) and the lower 8 bits loaded as data byte 2. This can be done as follows:

Upper byte = N >> 8
Lower byte = N & 0xFF

The third data byte specifies the search moe, mono-stereo operation, and left-right mute. The fourth data byte specifies the standby mode, band limits, clock freauency, noise cancelling etc. The fifth data byte PLL reference frequencyand the de-amphasis time constant.

Block diagram: Figure 4.72 shows the project block diagram, An earphone is connected to the output of the TEA5767 module.

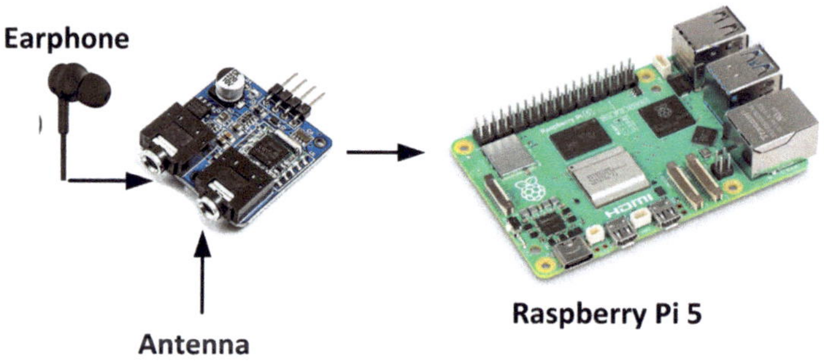

Figure 4.72: Block diagram of the project.

Circuit diagram: The circuit diagram is shown in Figure 4.73. SDA (GPIO2, pin 3) and SCL (GPIO3, pin 5) of Raspberry Pi 5 are connected to the radio module which is powered from the +5V of Raspberry Pi 5.

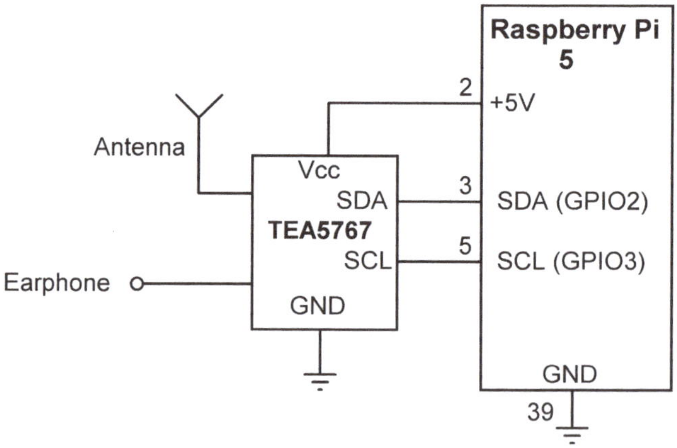

Figure 4.73: Circuit diagram of the project.

Before developing the program, you will have to enable I²C using the **raspi-config** utility. Then, construct the program as shown in Figure 4.73. Enter the following command to make sure that the TEA5767 module is detected by the Raspberry Pi 5 and read its address as shown in Figure 4.74. By default the address is 0x60.

```
pi@raspberrypi:~ $ i2cdetect -y 1
     0  1  2  3  4  5  6  7  8  9  a  b  c  d  e  f
00:                          -- -- -- -- -- -- -- --
10: -- -- -- -- -- -- -- -- -- -- -- -- -- -- -- --
20: -- -- -- -- -- -- -- -- -- -- -- -- -- -- -- --
30: -- -- -- -- -- -- -- -- -- -- -- -- -- -- -- --
40: -- -- -- -- -- -- -- -- -- -- -- -- -- -- -- --
50: -- -- -- -- -- -- -- -- -- -- -- -- -- -- -- --
60: 60 -- -- -- -- -- -- -- -- -- -- -- -- -- -- --
70: -- -- -- -- -- -- -- --
pi@raspberrypi:~ $
```

Figure 4.74: Fetch the I²C addess.

Program listing: Figure 4.75 shows the program listing. The program (Program: **fmradio.py**) consists of the following functions:

init(): This function initualizes the TEA5767 module.

set_freq(freq): This function receives the required station frequency as its parameter. The function loads the 5 data bytes to start receiving the broadcast at this frequency. At the beginning of the function, **N** is used to set the PLL value for the required frequency as shown in the above formula. **Nupper** and **Nlower** are the upper and lower bytes of **N** respectively. Notice that the mute bit (bit 7) is cleared so that the module is not in mute mode.

Array **buff** is used to hold the data bytes. **buff[0]** is loaded wih the lower part of the requires frequwncy, **buff[1]** is loaded with the following bits enabled (see the TEA5767 data sheet):

Bit	**Symbol**	**Definition**
7	SUD	enable search UP
6,5	SSL	search stop level low
4	HLSI	high side LO injection
0-3	MS,MR,ML,SWP1	mute is not enabled, port SWP1 is low

buff[2] and **buff[3]** are loaded such that (see the TEA5767 data sheet) the clock frequency is set to 32.768 kHz. The 5 byte **buff** data is then loaded to the chip using the I²C function **i2c.write_i2c_block_data**, with the upper part of the frequency loaded as the second argument of the call The set frequency is displayed on the screen.

Mute: this function mutes the module so that no broadcast is received and no output is heard. Here, the frequency is set to 0, and mute bit (bit 7) is set in **buff[0]**. The other bytes are left as in function **set_freq**. The message **Radio is muted...** is displayed on the screen.

Inside the main program, the frequency is initially set to 97.3 MHz which is the local radio station broadcasting frequency in London. The user is then prompted to enter a one character command. These commands are used either to select the prr-defined default frequencies or to increment/decrement the frequency. Valid commands and their meanings are as follows:

Command	**meaning**
f	start receiving from default local station at 97.3 MHz
n	start receiving from the next popular local station. Set to 105.4 MHz
+	increment the frequency by 1 MHz
	decrement the frequency by 1 MHz
>	increment the ferquency by 0.1 MHz
<	decrement the frequency by 0.1 MHz
m	mute
u	unmute
q	quit (mute and exit). Message **Exitting** is displayed

```
#---------------------------------------------------------------------
#                          FM RADIO
#                          ========
#
# This is an FM radio using the TEA5767 radio module. The module
# operates with the I2C signals SDA and SCL, connected to Raspberry
# PI 5 I2C pins.
#
# Author: Dogan Ibrahim
# File  : fmradio.py
# Date  : February, 2024
#----------------------------------------------------------------------
import smbus as smbus
i2c=smbus.SMBus(1)
DEVICE_ADDR = 0x60
import time

buff = [0]*4
print("TEA5767 FM RADIO")
print("================")

#
# TEA5767 Radio module connected to SDA0, SCL0
#

def init():
   i2c.write_quick(DEVICE_ADDR)
   time.sleep(0.1)

#
# This function sets the radio to required frequency
#
def set_freq(freq):
    N = int (4 * (freq * 1000000 + 225000) / 32768)
    Nupper = N >> 8
```

```
    Nlower = N  & 0xFF

    buff[0] = Nlower
    buff[1] = 0xB0
    buff[2] = 0x10
    buff[3] = 0x00

    i2c.write_i2c_block_data(DEVICE_ADDR, Nupper,buff)
    print("Frequency set to: " + str(freq))

#
# This function mutes the radio
#
def mute():
    N = int(4 * (0 * 1000000 + 225000) / 32768)
    Nlower = N & 0xFF
    init = 0x80
    buff[0] = Nlower
    buff[1] = 0xB0
    buff[2] = 0x10
    buff[3] = 0x00
    time.sleep(1)
    i2c.write_i2c_block_data(DEVICE_ADDR, init,buff)
    print("Radio Muted...")


init()
frequency = 97.3                                        # Starting frequency

try:
        while True:
            c = input("command (fn+-><muq): ")
            print(c)
            if c == 'f':                                # Set to 101.4
                frequency = 97.3
                set_freq(frequency)
                time.sleep(1)
            elif c == 'n':                              # Set to 102.0
                frequency = 105.4
                set_freq(frequency)
                time.sleep(1)
            elif c == '+':                              # Increment by 1 MHz
                frequency = frequency + 1
                set_freq(frequency)
                time.sleep(1)
            elif c == '-':                              # Decrement by 1 MHz
```

```
            frequency = frequency - 1
            set_freq(frequency)
            time.sleep(1)
        elif c == '>':                              # Increment by 0.1 MHz
            frequency = frequency + 0.1
            set_freq(frequency)
            time.sleep(1)
        elif c == '<':                              # Decrement by 0.1 MHz
            frequency -= 0.1
            set_freq(frequency)
            time.sleep(1)
        elif c == 'm':                              # Mute
            mute()
            time.sleep(1)
        elif c == 'u':                              # Unmute
            set_freq(frequency)
            time.sleep(1)
        elif c == 'q':                              # Mute and Exit
            mute()
            print("Exitting")
            break

except KeyboardInterrupt:
        print("Command error")
```

Figure 4.7: Program listing

Testing the radio

To check for proper operation, follow these steps.

- Construct the circuit.
- Connect the antenna and earphone to the module.
- Enter the following command to start the program:

  pi@raspberrypi:~ $ **python fmradio.py**

- Enter **f** and then Enter. By default, the radio will receive broadcast from 97.3 MHz (change this if required). Use the commands to increase/decrease the frequency as required. You should receive the broadcasts on your earphone. Type **q** followed by the **Enter** key to terminate.

Figure 4.76 shows an example run of the program.

```
pi@raspberrypi:~ $ python fmradio.py
TEA5767 FM RADIO
================
command (fn+-><muq): f
f
Frequency set to: 97.3
command (fn+-><muq): -
-
Frequency set to: 96.3
command (fn+-><muq): -
-
Frequency set to: 95.3
command (fn+-><muq): +
+
Frequency set to: 96.3
command (fn+-><muq): +
+
Frequency set to: 97.3
command (fn+-><muq): >
>
```

Figure 4.76: Example run of the program.

Note: Make sure that the TEA5767 is as far as possible from the Raspberry Pi 5 board as repetitive clicking noise may occur in the audio output when the two are close to each other.

4.13 Project 12: Modified Project — Increasing the Output Signal Level - Connecting a Loudspeaker

Description: In the previous project, you had to use earphones to listen to your radio station(s). An audio power amplifier can be used to increase the signal level so that the radio can be connected to a loudspeaker.

There are many low-cost audio power amplifier modules available in the market. One of the cheapest options is to use the type LM386 audio amplifier IC. The disadvantages of the LM386 are that it is (1) mono and (2) its output power may be too low, requiring further amplification. In this section you will see how to connect an LM386 module to your radio.

Figure 4.77 shows the LM386 amplifier module. The module consists of an LM386 chip, a potentiometer to adjust the volume, a capacitor, a few resistors, a 4-way connector to hook up the audio input and the power supply, and a 2-way screw terminal connector for a small loudspeaker.

Figure 4.77: The LM386 audio amplifier module.

Block diagram: The block diagram of the modified project is pictured in Figure 4.78.

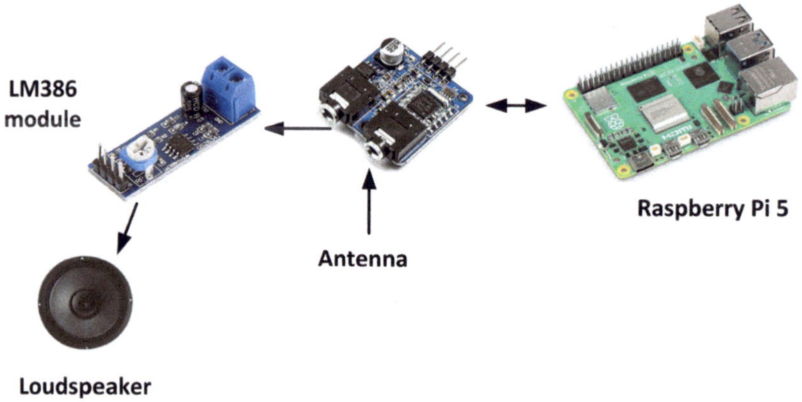

Figure 4.78: Block diagram of the modified project.

Circuit diagram: The modified circuit diagram is shown in Figure 4.79. The amplifier module is operated at +5 V supply voltage taken from the Raspberry Pi 5.

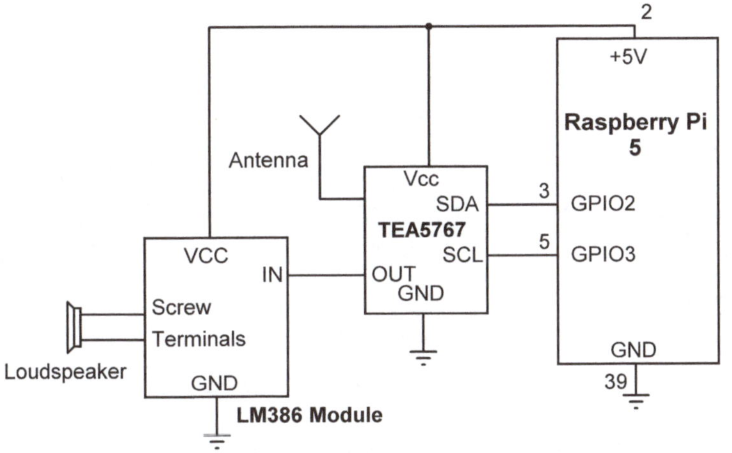

Figure 4.79: Circuit diagram of the modified project.

The author recommends using a more powerful stereo audio amplifier module. Some models which are low-cost and available on Amazon are shown in Figure 4.80. Interested amateur radio readers may prefer to design and construct transistor-based audio amp circuits. A volume control can also be added to the project with the audio amplifier. Alternatively you can use the speaker+amplifier module that you will be using in your RTL-SDR projects (see Chapter 5).

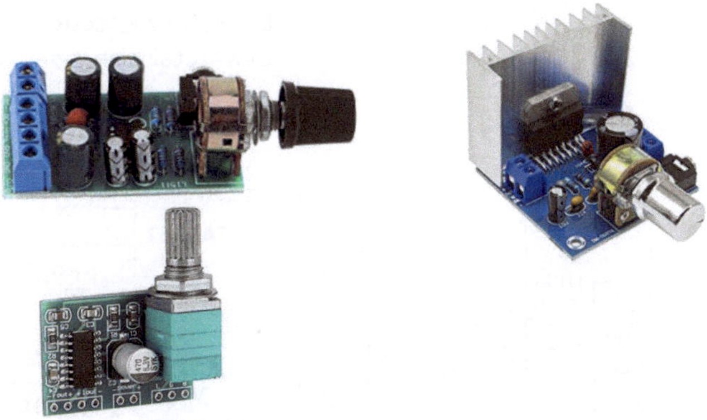

Figure 4.80: Some low-cost audio power amplifier modules.

4.14 Project 13: FM Radio using an LCD and Rotary Encoder to Set the Frequency

Description: This project is similar to the previous one but here a rotary encoder is used to set the frequency instead of setting it from the keyboard. Setting the frequency with a rotary encoder is accurate and very easy. In this project, turning the encoder shaft by one click changes the frequency by 100 kHz but obviously that can be changed if desired.

Block diagram: Figure 4.81 shows the block diagram of the project. The frequency settings are displayed on the LCD as the rotary encoder is turned.

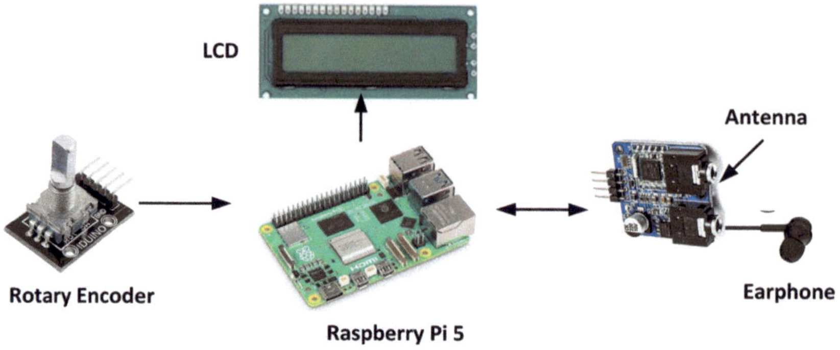

Figure 4.81: Block diagram of the project.

Circuit diagram: The circuit diagram of the project is shown in Figure 4.82. The connections between the Raspberry Pi 5 and the external components are as follows. Notice that the I²C LCD and the TEA5767 both use the same I²C port of the Raspberry Pi 5 and because they have different addresses, there is no problem:

Raspberry Pi 5 GPIO	**Pin number**	**Connected to**
GPIO2	3	I²C LCD SDA, TEA5767 SDA
GPIO3	5	I²C LCD SCL, TEA5767 SCL
GPIO21	40	SW Rotary encoder
GPIO20	38	DT Rotary encoder
GPIO16	36	CLK Rotary encoder

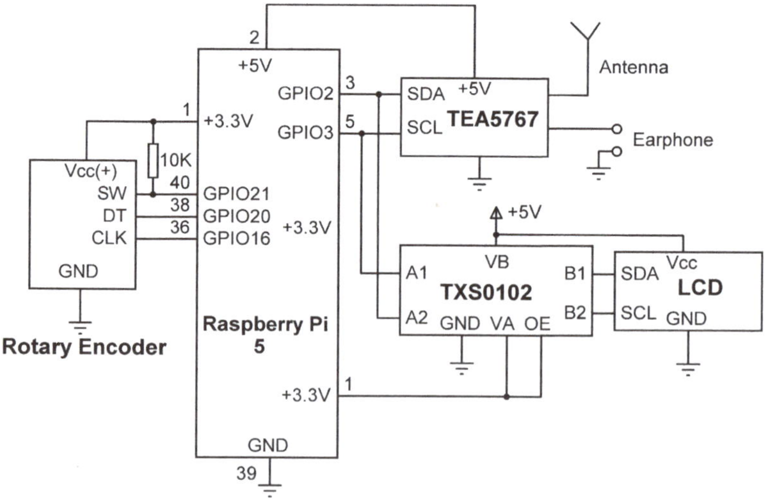

Figure 4.82: Circuit diagram of the project.

Notice that a voltage level converter module is used for the I²C LCD connections.

Program listing: Figure 4.83 shows the program listing (Program: **RadioRotary.py**). At the beginning of the program, the modules used are imported, I²C LCD connection is defined, and text **TEA5767 FM RADIO** is displayed on the top row of the LCD. Then, the rotary encoder connections and the TEA5767 FM radio module connections are defined. Functions **init()** and **set_freq()** are as in the previous projects. The main program is executed in a loop forever using a **while** statement. Inside this loop, the rotary encoder turnings are sensed and the frequency is changed by 0.1 MHz at each click of the rotary encoder (this value can be changed if desired). The resulting frequency is sent to function **set_freq()** which then configures the radio module to be set to this frequency and receive broadcasts. Make sure that you connect the antenna and an earphone to the TEA5767 radio module before testing the project. Pushing in the encoder knob terminates the program.

```
#-------------------------------------------------------------------
#                  FM RADIO WITH LCD AND ROTARY ENCODER
#                  ====================================
#
# This is an FM radio using the TEA5767 radio module. The module
# operates with the I2C signals SDA and SCL, connected to Raspberry
# PI 5 I2C pins. Additionally, a rotary encoder and an LCD are
# connected to the circuit. The rotary encoder is used to increase
# or decrease the frequency in steps of 0.1 MHz. It is recommended
# to use an audio amplifier to increase the audio signal level.
#
# Author: Dogan Ibrahim
# File  : RadioRotary.py
# Date  : February, 2024
#---------------------------------------------------------------------
from gpiozero import InputDevice

#
# Import I2C libraries
#
import smbus as smbus
i2c=smbus.SMBus(1)
DEVICE_ADDR = 0x60

import time

#
# Import LCD libraries
#
from lcd_api import LcdApi
from i2c_lcd import I2cLcd

I2C_ADDR = 0x27
I2C_NUM_ROWS = 2
```

```
I2C_NUM_COLS = 16

mylcd = I2cLcd(1,I2C_ADDR, I2C_NUM_ROWS, I2C_NUM_COLS)
mylcd.clear()
mylcd.move_to(0,0)
mylcd.putstr("TEA5767 FM RADIO")
time.sleep(3)

buff = [0]*4

#
# Rotary encoder connections
#
SW = 21
DT = 20
CLK = 16

#
# Rotary encoder configuration
#
clk =  InputDevice(CLK)
dt = InputDevice(DT)
sw = InputDevice(SW)

#
# TEA5767 Radio module connected to SDA, SCL
#

def init():
   i2c.write_quick(DEVICE_ADDR)
   time.sleep(0.1)

#
# This function sets the radio to required frequency
#
def set_freq(freq):
    N = int (4 * (freq * 1000000 + 225000) / 32768)
    Nupper = N >> 8
    Nlower = N  & 0xFF

    buff[0] = Nlower
    buff[1] = 0xB0
    buff[2] = 0x10
    buff[3] = 0x00

    i2c.write_i2c_block_data(DEVICE_ADDR, Nupper,buff)
```

```
    mylcd.clear()
    mylcd.move_to(0,0)
    mylcd.putstr("freq set to:")
    mylcd.move_to(0,1)
    mylcd.putstr(str(freq)[:6]+" MHz")

#
# This function mutes the radio
#
def mute():
    N = int(4 * (0 * 1000000 + 225000) / 32768)
    Nlower = N & 0xFF
    init = 0x80
    buff[0] = Nlower
    buff[1] = 0xB0
    buff[2] = 0x10
    buff[3] = 0x00
    time.sleep(1)
    i2c.write_i2c_block_data(DEVICE_ADDR, init,buff)
    mylcd.clear()
    mylcd.move_to(0,0)
    mylcd.putstr("Exit")

init()
frequency = 97.3                                        # Starting frequency
set_freq(frequency)

flag = 1
ClkOldState = clk.value

#
# Get the required frequency. Each click of the rotary encodes
# increment or decrements the frequency by 0.1 MHz
#
while flag == 1:
   ClkState = clk.value
   DTState = dt.value
   if ClkState != ClkOldState and ClkState == 1:
      if DTState != ClkState:
            frequency = frequency + 0.1
      else:
            frequency = frequency - 0.1
      set_freq(frequency)
   ClkOldState = ClkState

   SWState = sw.value
```

```
    if SWState == 0:                        # If SW pressed
        flag = 0
        mute()                              # Mute and exit
        time.sleep(1)
```

Figure 4.83 Program listing

Note: Make sure that the TEA5767 is as far as possible from the Raspberry Pi 5 board as repetitive clicking noise may occur in the audio output when the two are close to each other.

Complete Raspberry Pi 5 FM radio

1. Battery operation
Now that the project is independent of the PC, you might want to operate it with a battery. Perhaps the easiest option is to use a power bank used to charge mobile phones. Then, the Raspberry Pi 5 can be connected to the power pack using the micro USB cable.

2. Auto boot
When operating independent of the PC, you may want the radio program to start as soon as power is applied to the Raspberry Pi 5. The auto boot mechanism is explained earlier in this Chapter.

3. Using an audio amplifier module
As described in the previous project, an audio amplifier module will be required before the radio can be connected to a loudspeaker. As an example, you can use the audio amplifier shown in Figure 4.84. This amplifier conveniently has built-in volume control.

Figure 4.84: Audio amplifier module.

4. Radio case front panel layout
A suitable front panel layout for your radio is shown in Figure 4.85. A block diagram of the components inside the case is shown in Figure 4.86.

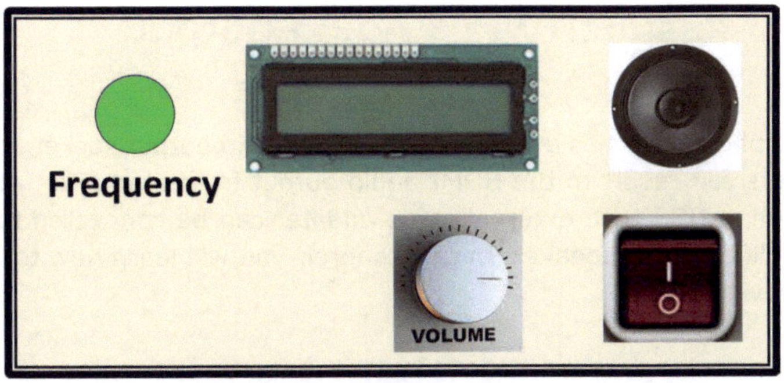

Figure 4.85: Suggested front panel layout.

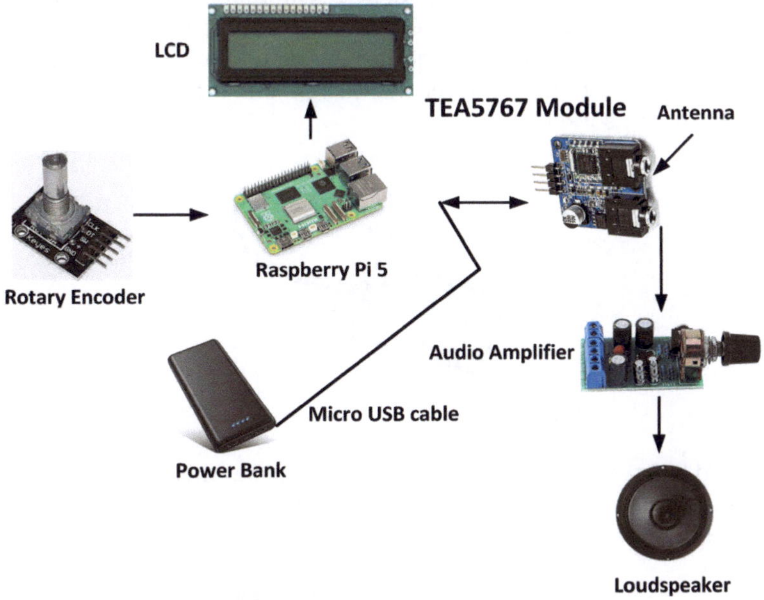

Figure 4.86: Block diagram of the radio.

Chapter 5 • Raspberry Pi 5 Audio Output

5.1 Overview

Whereas the Raspberry Pi 4 has a 3.5-mm jack type audio output socket, the Raspberry Pi 5 has none. Users can revert to the HDMI audio output from a monitor. Alternatively and preferably though, a USB type external audio adapter can be connected to the Raspberry Pi 5 plus an amplifier based speaker. In this chapter, you will learn how to use such a USB audio adapter device.

5.2 Using an External USB Audio Adapter

In many radio amateur applications, you will need audio output to a speaker or headphones. The authors recommend the use of a USB audio adapter like the UGREEN (see Figure 5.1), or one with a built-in amplifier and volume control. Having a hardware volume control has the advantage that the volume can be set easily. The authors employ the UGREEN audio adapter which has two sockets — one for the microphone input (pink socket) and one for the audio output (green socket).

Figure 5.1: UGREEN audio adapter.

You will also need to have a speaker with a built-in amplifier module, having a 3.5-mm jack plug and preferably powered by USB. An example speaker is shown in Figure 5.2, although there are many such speakers available in the market.

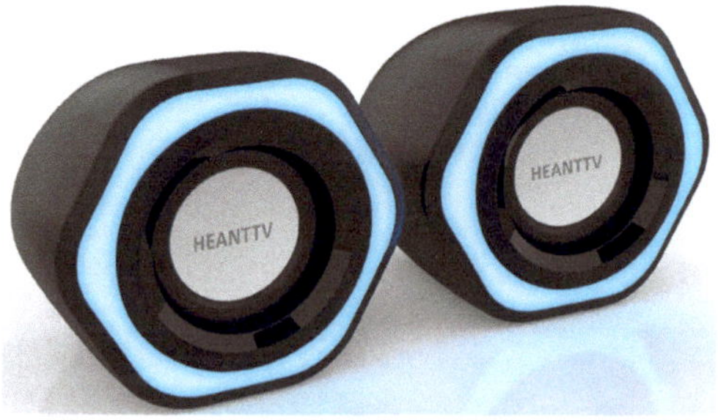

Figure 5.2: Speaker with built-in amplifier.

5.3 Testing the Audio Output

Connect the USB audio adapter to one of the USB ports of your Raspberry Pi 5 and connect the speaker and amplifier module to the green port of the UGREEN adapter. Enter the following command to list the audio devices available on your system:

pi@raspberrypi:~ $ **aplay –l**

Notice that as shown in Figure 5.3, the USB audio adapter is on **card: 2**.

```
pi@raspberrypi:~ $ aplay -l
**** List of PLAYBACK Hardware Devices ****
card 0: vc4hdmi0 [vc4-hdmi-0], device 0: MAI PCM i2s-hifi-0 [MAI PCM i2s-hifi-0]
  Subdevices: 1/1
  Subdevice #0: subdevice #0
card 1: vc4hdmi1 [vc4-hdmi-1], device 0: MAI PCM i2s-hifi-0 [MAI PCM i2s-hifi-0]
  Subdevices: 1/1
  Subdevice #0: subdevice #0
card 2: Device [USB Audio Device], device 0: USB Audio [USB Audio]
  Subdevices: 1/1
  Subdevice #0: subdevice #0
```

Figure 5.3: Audio devices.

Enter the following command to test the audio output. This test will generate a 500 Hz sine wave signal on the speaker:

pi@raspberrypi:~ $ **speaker-test –c2 –t sine –f 500**

You can also download a short **.wav** type audio file (e.g. music file) to your Raspberry Pi 5 home folder and then run the file to hear the music. For example, the authors downloaded and tested the music file named: **sample-file-4.wav**:

pi@raspberrypi:~ $ **aplay sample-file-4.wav**

Press Cntrl+C to stop the music.

5.4 Audio Volume

Most speakers with built-in amplifiers have volume control buttons that you can adjust the volume output. You may also use the **alsamixer** program on your Raspberry Pi 5 to set the volume in software. The steps are as follows.

- Enter the following command and you will see the screen as in Figure 5.4:

    pi@raspberrypi:~ $ **alsamixer**

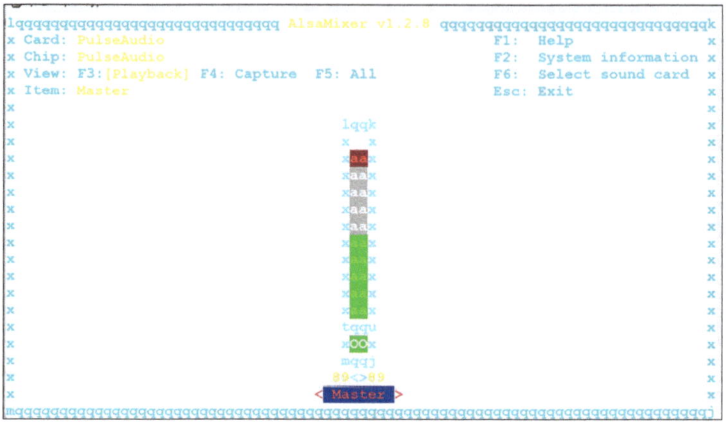

Figure 5.4: The alsamixer screen.

- Press the F6 function key and select the **USB audio device** (Figure 5.5), then press the **Enter** key

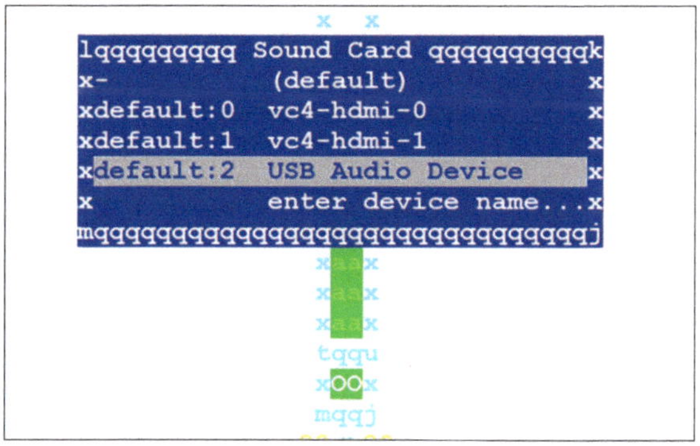

Figure 5.5: Select the USB audio device.

- Click on the column **Speaker** and use the up-down arrow keys to increase/decrease the speaker volume. Figure 5.6 shows the maximum volume setting.

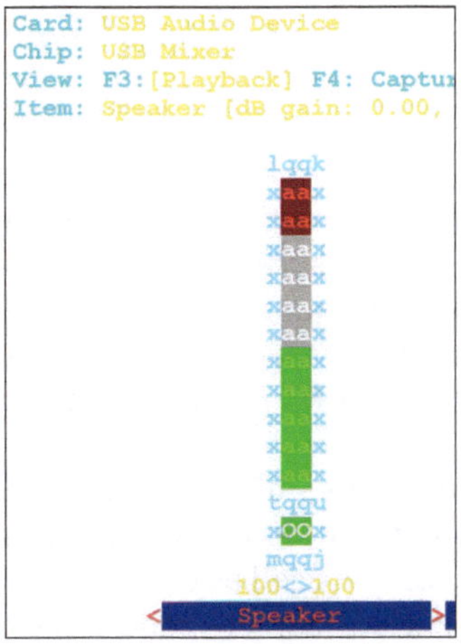

Figure 5.6: Maximum speaker volume setting.

- Press the **ESC** key to exit from the **alsamixer** program.

Chapter 6 • RTL-SDR Meets Raspberry Pi 5

6.1 Overview

RTL-SDR is a cheap USB dongle that can be used as a computer based radio scanner for receiving live radio signals without Internet access. RTL-SDR is normally used as a receiver only and unable to transmit radio signals. Depending on the particular model you purchase, you can receive radio signals in the frequency range from 500 kHz to 1.75 GHz. There are also other more expensive software defined radio modules upping on the RTL-SDR, some of which can even transmit (e.g. HackRF). All of these all come at a higher price though.

In this chapter, you will explore the use of your Raspberry Pi 5 in conjunction with RTL-SDR dongles. Most RTL-SDR dongles in the market and based on the RTL2832U chip have recently become popular as they can be used as cheap SDRs. There are many different types of RTL-SDR dongles in the market but earlier models are mostly based on the RTL2832U chip, while the newer dongles have the R828D chip. There are currently two versions of the dongles: V3 and V4. Figure 6.1 shows pictures of both types of dongles.

Figure 6.1: V3 and V4 RTL-SDR modules.

Back in 2012, an undocumented feature of the RTL2832U chip surfaced that enabled the device to be used as a general-purpose SDR. With the development of hardware and software, these cheap dongles can now be used as sophisticated SDR receivers. For example, an RTL-SDR dongle can be connected to a Raspberry Pi and many interesting amateur radio software packages can be downloaded and cheerfully operated on Raspberry Pi.

Without these dongles, such receivers would have cost hundreds or thousands of dollars to bring similar features. Application areas of RTL-SDR dongles include:

- Listening to air traffic control conversations
- Tracking aircraft positions
- Receiving meteorological transmissions
- Listening to amateur radio bands
- Listening to shortwave and FM radio
- Monitoring meteor scatter
- Listening to DAB broadcast transmissions
- Listening to International Space Station
- Listening to amateur radio
- Receiving HF weatherfax
- Listening short wave radio
- Receiving and decoding GPS signals
- Decoding ham radio APRS packets

- Developing a radio scanner
- Watching analog broadcast TV
- Using it as a spectrum analyzer
- Using it as a receiver server
- Receiving GPS signals and decoding them
- Scanning cordless phones and baby monitors
- Watching weather satellite
- Radio astronomy
- and more

The minimum requirements to use an RTL-SDR dongle are:

- an RTL-SDR dongle device;
- a suitable antenna;
- a powerful computer (like Raspberry Pi 5);
- RTL-SDR driver and application software (there are many and most are free).

Depending on your requirements and applications, you may also need to use filters to improve the signal to noise ratio.

6.2 RTL-SDR V3

Earlier RTL-SDR V3 devices could work from about 500 kHz to over 1.7 GHz. The lower frequency range can be extended by using an **upconverter** device or by **direct sampling mode**. The upconverter is connected to the antenna ahead of the RTL-SDR device. Some upconverters operate down to several kHz. If the upconverter oscillator frequency is 125 MHz and you want to tune to 5 MHz, simply tune your receiver (or GQRX, for example) to 130 MHz. The direct-sampling mode requires a small change to be made to the RTL-SDR hardware and the software, where a connection is made inside the hardware. Some RTL-SDR devices have a small hole allowing a connection to be made by inserting a jumper wire, thus there is no need to open the device and get the solder iron out. The software should be configured so that the sampling mode is set to Direct Sampling.

The RTL-SDR has 3.2 MHz bandwidth (2.4 MHz stable), 8-bit analog-to-digital converters (ADC), less than 4.5 dB noise figure, and 75 ohms input impedance (not 50 ohms which is the *de facto* impedance used in amateur radio. In general, the mismatch between the 75 and 50 ohms is less than 0.2 dB). Because the RTL-SDR devices are cheap, they use 28.8 MHz crystal oscillators and their clock accuracy may drift several kHz. Most popular RTL-SDR software packages have options in the form of ppm drift values for calibrating this drift in software. There are also noise bursts in the form of harmonics at multiples of 28.8 MHz. These bursts or "spikes" can usually be observed in waterfall displays.

For good reception and low noise, the RTL-SDR should be placed close to the antenna so that the lossy coaxial cable is replaced with loss-free USB cable. You should take care however that the length of the USB cable is not more than 5 meters for USB2.0, or 3 meters for USB3.0. If longer length USB cables are required, then it is recommended to use USB hubs or USB repeater devices.

The RTL-SDR dongles should be installed in metal enclosures in order to minimize external interferences and they should not be near power lines, motors, TVs, appliances, or other equipment likely to generate electromagnetic noise.

Having a good antenna is also very important while using your RTL-SDR. The cheap antenna typically delivered with the device is about 12-15cm long and not generally suitable for amateur work. You should place this antenna over a metal surface, which has a radius of around 12-15 cm minimum to make a quarter-wave "Marconi" antenna and improve the reception characteristics.

6.3 RTL-SDR V4 vs V3

The new V4 dongle has many improvements compared to the earlier V3 dongle:

- V4 has built-in upconverter that improves the HF reception and provides adjustable gain on HF.
- V4 provides improves filtering as it uses the R828D tuner chip with inputs for three bands HF, VHF, and UHF. As a result of this out of band interference from strong broadcasting stations is less likely
- In addition to improved filtering discussed above, the R828D chip in the V4 version has an open-drain pin which allows simple notch filters to be added to minimize interference mainly from the AM broadcasts.
- The power supply noise on V4 is significantly reduced due to improved power supply design.
- The V4 dongle uses less current and consequently generates less heat compared to the V3.
- The price tag of the V4 was initially higher than that of the V3 but is currently at level.

Both the V3 and V4 are built with black metal enclosures, which act as a heatsink for passive cooling and also protect from electromagnetic fields.

V4, however, has some disadvantages compared to V3:

- Reduced sensitivity on some bands due to increased filtering
- Requires different drivers than V3
- The R828D tuner chip is currently out of production and it may be difficult to obtain V4 dongles.

Figures 6.2 and 6.3 show the inside of V3 and V4 dongles, respectively.

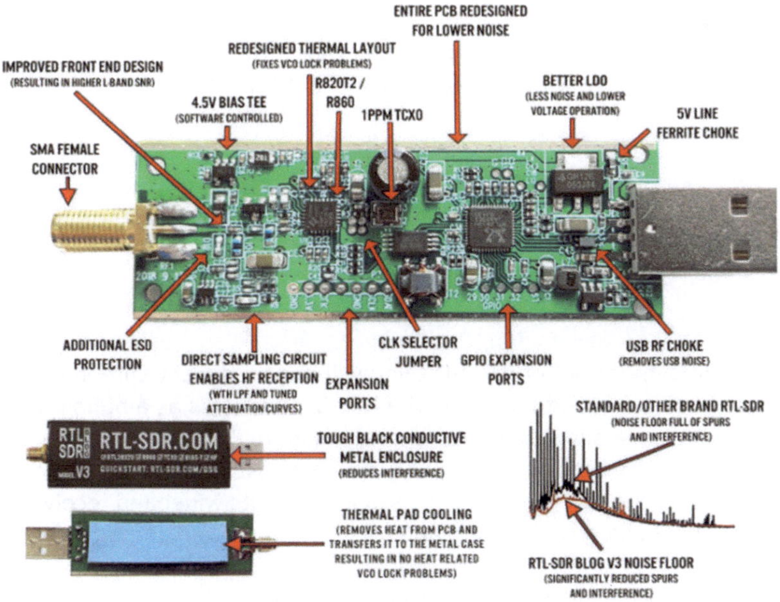

Figure 6.2: Inside the V3 dongle.

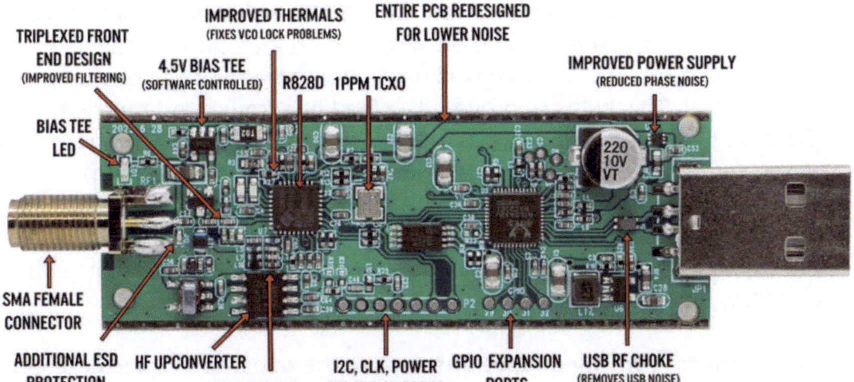

Figure 6.3: Inside the V4 dongle.

Source: https://www.rtl-sdr.com/rtl-sdr-blog-v4-dongle-initial-release/

It is interesting to note that the RTL-SDR is community based and is not owned by any company or person. It is all community supported open software and hardware. It came into life by combined efforts of Antti Palosaari, Eric Fry and Osmocom (in particular Steve Markgraf). It was first discovered that certain TV dongles could be used for SDR. Osmocom in particular developed the first RTL-SDR driver which was released as open-source.

In this chapter, we are only interested in using an RTL-SDR device with the Raspberry Pi 5. Although the V4 dongle is used in the book, the driver is downward compatible and will work with V3 dongles as well. The theory and operation of the RTL-SDR devices are beyond the scope of this book. Interested readers can get tons of information on RTL-SDR from the Internet and from many books available on this topic. The following websites could be of interest to readers who may want to learn more about the RTL-SDRs:

https://osmocom.org/projects/rtl-sdr/wiki/Rtl-sdr
https://www.elektor.com/elektor-raspberry-pi-rtl-sdr-v4-bundle-en
https://www.rtl-sdr.com/V4/
https://www.rtl-sdr.com/rtl-sdr-quick-start-guide/
https://www.rtl-sdr.com/using-our-new-dipole-antenna-kit/

6.4 The RTL-SDR Antenna Kit

To get the most enjoyment out of RTL-SDR you will need a decent antenna. Some retailers and distributors (including Elektor) supply the RTL-SDR dongles as a bundle with antennas. A dipole antenna kit is usually supplied as shown in Figure 6.4. Be sure to get the dipole up high and outside (during good weather only) for best results. The recommended outdoor antenna for general scanning is a so-called discone due to its wideband receiving properties. You can also cheaply build a wideband planar disk antenna from some metal pizza pans.

This dipole antenna kit includes the following (see Elektor site: https://www.rtl-sdr.com/using-our-new-dipole-antenna-kit/):

- **1× dipole antenna base with 60 cm RG174 cable and SMA Male connector.** This is the dipole base where the telescopic antennas connect to. The short run of RG174 coax is decoupled from the base elements with a ferrite choke. This helps to prevent the feed line from interfering with the dipole radiation pattern. The dipole has a 1/4 inch threading on the bottom, which allows you to use standard camera mount products for mounting.

- **1× 3 meter RG174 coax cable extension.** This coax cable extension allows you to mount the antenna in a place that gets better reception, like outside on a window, or higher up.

- **2× 23 cm to 1 m telescopic antennas.** The telescopic dipoles are detachable from the dipole base via an M5 thread, which allows for greater portability and the ability to swap them out. These long telescopic antennas cover VHF to UHF.

- **2x 5 cm to 13 cm telescopic antennas.** These smaller antennas cover UHF to 1090 MHz ADS-B, and even still work decently up to L-band 1.5 GHz frequencies.

- **1× flexible tripod mount with 1/4" screw.** This piece allows you to mount the dipole on a variety of different locations like on a pole, a tree branch, a desk, a door, or a window sill. The legs of the tripod are bendy and rubberized so can wrap securely around many objects.

- **1× suction cup mount with 1/4" screw.** With this mount, you can install the dipole on the outside of a window, on a wall, car roof/window, or on any other smooth surface. To use it, first clean the surface with window cleaner or isopropyl alcohol. Then place the suction cup on the cleaned surface and close the lever to activate the suction.

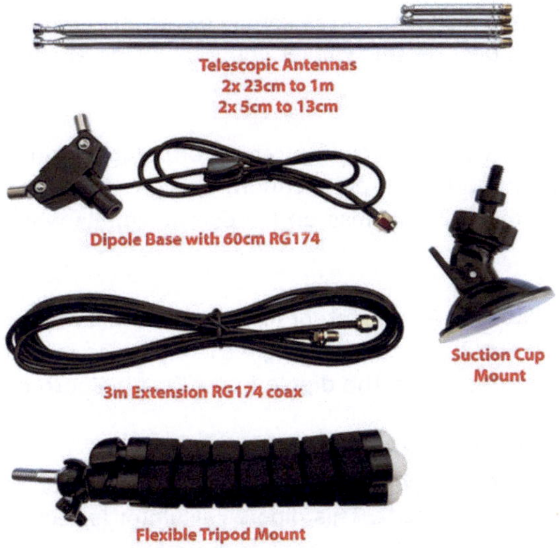

Figure 6.4: RTL-SDR dipole antenna kit.

6.4.1. Dipole orientation

Signals are normally transmitted with either horizontal, vertical or right hand/left hand circular polarization (RHCP/LHCP). This is essentially the "orientation" of a signal. At the receiver side, an antenna with the same polarization should be used too for best performance. A dipole can be used in either vertical or horizontal polarization just by orienting it either vertically or horizontally.

If you mismatch vertical and horizontal polarization or RHCP and LHCP, you'll get an instant 20 dB loss. If you mismatch vertical/RHCP, vertical/LHCP, horizontal/RHCP, horizontal/LHCP, you'll only get a 3 dB loss.

For vertical polarization, in theory it does not matter which way around you orient the antenna as long as it is vertical. However, in practice, you may get slightly better results by having the element connected to the center coax conductor pointing UP. You can confirm which element is connected to the center conductor by temporarily removing the black lid on the dipole base (it can be easily pried off with a nail or flat head screwdriver).

There are also ways to optimize the radiation pattern with dipoles. For example, for LEO VHF satellites, you can use a V-dipole configuration. You can also make a somewhat directional antenna by using a bent dipole configuration.

6.4.2. Terrestrial signal reception

Most signals broadcast terrestrially (on Earth) are vertically polarized. To use the dipole for vertically polarized signals, all that you need to do is orient the elements vertically (up and down). In theory there is no such thing as up and down for the dipole when used in the vertical orientation. However, in practice you may find slightly better performance when the "active" element points up. The active element is the one connected to the center conductor. You can check which element is connected to the center conductor by removing the top cap on the dipole base — this will let you look inside at the connections.

6.4.3. Satellite reception

The dipole can be used in a V-Dipole configuration for polar orbiting satellite reception. The idea is to use the dipole in horizontal polarization. This gives 3 dB loss on the RHCP satellite signals, but also nicely gives 20 dB loss on terrestrial signals which could be overloading your RTL-SDR.

For 137 MHz satellites like NOAA and Meteor M2, extend the larger antenna elements out to about 53.4 cm each (about 2.5 sections). Angle the dipole so it is horizontal and in a 'Vee' shape, at about 120 degrees. Place the dipole in the North-South direction.

6.4.4. Choosing the antenna element length

Just as with the whip antenna, you can use an online calculator to calculate the optimal length for your frequency of interest. This dipole calculator is recommended:

http://www.csgnetwork.com/antennaedcalc.html

The exact length does not matter too much, but try to get the lengths as close as you can to what the calculator says. With the pure dipole, you want both elements to be the same length. In reality, extending the antenna to almost any random length will work just fine for most strong signals. But if you're really trying to optimize those weak signals you'll want to fine tune the lengths.

Basically, the longer the antenna, the lower its resonant frequency. The shorter the antenna, the higher the resonant frequency. You want to be close to the resonant frequency. Remember that there is about 2 cm of metal inside the antenna itself, which needs to be added on. Below is a cheat sheet for various lengths and frequencies. Note that the length refers to the length of one side of the dipole only (e.g. the length that you need to extend each element out to).

- Large Antenna, 5 Sections, 100 cm + 2 cm is resonant @ ~70 MHz
- Large Antenna, 4 Sections, 80 cm + 2 cm is resonant @ ~87MHz
- Large Antenna, 3 Sections, 60 cm + 2 cm is resonant @ ~115 MHz
- Large Antenna, 2 Sections, 42 cm + 2 cm is resonant @ ~162 MHz
- Large Antenna, 1 Section, 23 cm + 2 cm is resonant @ ~ 285 MHz
- Small Antenna, 4 Sections, 14 cm + 2 cm is resonant @ ~445 MHz
- Small Antenna, 3 Sections, 11 cm + 2 cm is resonant @ ~550 MHz
- Small Antenna, 2 Sections, 8 cm + 2 cm is resonant @ ~720MHz
- Small Antenna, 1 Section, 5 cm + 2 cm is resonant @ ~1030 MHz.

See the SWR plots at the end for a more accurate reading of the resonance points. But in most cases, no matter what you extend the length to, the SWR should be below 5 at most frequencies which results in 2.5 dB loss or less.

There are many online tools for calculating the antenna length for dipole, inverted-V, vertical antennas etc. Some interesting sites are:

https://m0ukd.com/calculators/quarter-wave-ground-plane-antenna-calculator/
https://www.westmountainradio.com/antenna_calculator.php
https://www.hamradiosecrets.com/antenna-calculator.html
https://www.qsl.net/w/w7lk/site/antennachart.htm
https://www.ws6x.com/ant_calc.htm#:~:text=The%20basic%20formula%20for%20determining,)%20%3D%20Length%20(feet).

6.5 Hardware Setup

Figure 6.5 shows the RTL-SDR hardware setup using a V4 dongle, dipole antenna kit, Raspberry Pi 5 computer, UGREEN audio adapter, and speaker with amplifier module. This is probably the minimum hardware setup you will require for your RTL-SDR based projects.

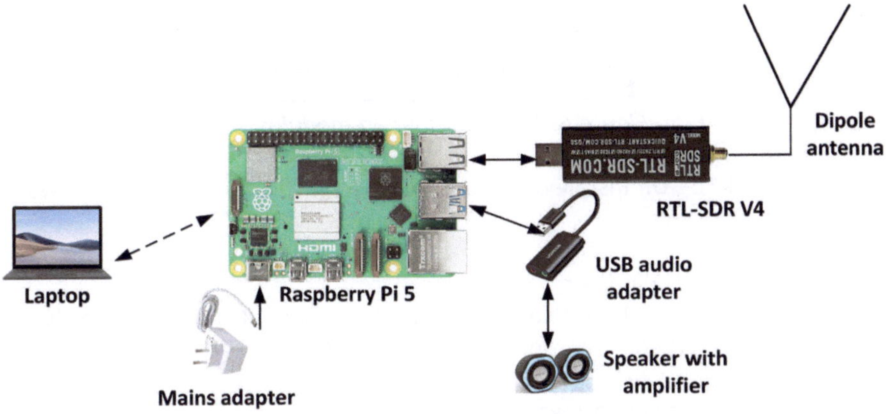

Figure 6.5: Hardware setup.

6.6 Installing the RTL-SDR Software on Raspberry Pi 5

Set up the hardware as in Figure 6.5 and enter the following command to list the recognized USB devices. Figure 6.6 shows the results where the RTL-SDR dongle is recognized as: **RealTek Semiconductor Corp. RTL2838 DVB-T**

pi@raspberrypi:~ $ **lsusb**

```
pi@raspberrypi:~ $ lsusb
Bus 004 Device 001: ID 1d6b:0003 Linux Foundation 3.0 root hub
Bus 003 Device 003: ID 0bda:2838 Realtek Semiconductor Corp. RTL2838 DVB-T
Bus 003 Device 002: ID 0d8c:0014 C-Media Electronics, Inc. Audio Adapter (Unitek
 Y-247A)
Bus 003 Device 001: ID 1d6b:0002 Linux Foundation 2.0 root hub
Bus 002 Device 001: ID 1d6b:0003 Linux Foundation 3.0 root hub
Bus 001 Device 001: ID 1d6b:0002 Linux Foundation 2.0 root hub
pi@raspberrypi:~ $
```

Figure 6.6: List of USB devices.

Although the dongle has been detected, it has been recognized as a DVB-T television dongle. Check to see what modules are loaded into the Kernel for the dongle. Enter (see Figure 6.7):

pi@raspberrypi:~ $ **lsmod | egrep 'sdr|dvb'**

```
pi@raspberrypi:~ $ lsmod | egrep 'sdr|dvb'
rtl2832_sdr              65536  0
videobuf2_vmalloc        65536  1 rtl2832_sdr
dvb_usb_rtl28xxu         65536  1
dvb_usb_v2               65536  1 dvb_usb_rtl28xxu
dvb_core                180224  2 dvb_usb_v2,rtl2832
videobuf2_v4l2           65536  4 rtl2832_sdr,pisp_be,rpivid_hevc,v4l2_mem2mem
videobuf2_common        114688  8 rtl2832_sdr,videobuf2_vmalloc,pisp_be,videobu
dma_contig,videobuf2_v4l2,rpivid_hevc,v4l2_mem2mem,videobuf2_memops
videodev                311296  6 rtl2832_sdr,pisp_be,videobuf2_v4l2,videobuf2_
mon,rpivid_hevc,v4l2_mem2mem
mc                       98304  9 videodev,snd_usb_audio,dvb_usb_v2,pisp_be,vid
uf2_v4l2,dvb_core,videobuf2_common,rpivid_hevc,v4l2_mem2mem
pi@raspberrypi:~ $
```

Figure 6.7: Kernel-loaded modules.

If you already have RTL-SDR installed on your Raspberry Pi 5, remove it as follows (ignore any error messages):

- sudo apt purge ^librtlsdr
- sudo rm -rvf /usr/lib/librtlsdr* /usr/include/rtl-sdr* /usr/local/lib/librtlsdr* /usr/local/include/rtl-sdr* /usr/local/include/rtl_* /usr/local/bin/rtl_*

Install the RTL-SDR drivers as follows:

- sudo apt-get install libusb-1.0-0-dev git cmake pkg-config
- git clone https://github.com/rtlsdrblog/rtl-sdr-blog
- cd rtl-sdr-blog
- mkdir build
- cd build
- cmake ../ -DINSTALL_UDEV_RULES=ON
- make
- sudo make install
- sudo cp ../rtl-sdr.rules /etc/udev/rules.d/
- sudo ldconfig
- sudo apt install debhelper
- cd rtl-sdr-blog
- sudo dpkg-buildpackage -b --no-sign

- cd ..
- sudo dpkg -i librtlsdr0_*
- sudo dpkg -i librtlsdr-dev_*
- sudo dpkg -i rtl-sdr_*

As you can see from Figure 6.6, you have to disable module: **dvb_usb_rtl28xxu** so that the dongle is not recognized as a TV device. The easiest option is to blacklist the dongle device so that it is not recognized as a TV dongle. The command is:

```
echo 'blacklist dvb_usb_rtl28xxu' | sudo tee --append /etc/modprobe.d/blacklist-dvb_usb_rtl28xxu.conf
```

Now reboot you Raspberry Pi 5:

```
sudo reboot
```

Enter the following command to check the modules loaded for the sdr/dvb. There should be no output displayed:

```
lsmod | egrep 'sdr|dvb'
```

Note: You must install the RTL-SDR software on your Raspberry Pi 5 before any other RTL-SDR based program is installed.

6.7 Testing — Tuning to a Frequency Manually

You can use the RTL-SDR software to tune to a frequency by entering a command at the command line. As an example, the following command tunes to frequency 97.3 MHz, which happens to be the LBC radio station broadcasting on FM in London:

```
pi@raspberrypi:~ $ rtl_fm –M wbfm –f 97300000 | aplay –r 32000 –f S16_LE –c 1
```

You should be able to hear the radio broadcast on the 97.3 MHz or the frequency you entered. Figure 6.8 shows the what is displayed by Raspberry Pi 5 when the above command is entered. Enter **Cntrl+C** to terminate the command and stop receiving the radio broadcast. If you are using the RTL-SDR V3, make sure that you connect it to a USB 2.0 port of your Raspberry Pi 5.

```
pi@raspberrypi:~ $ rtl_fm -M wbfm -f97300000 | aplay -r 32000 -fS16_LE -c 1
Found 1 device(s):
  0:  RTLSDRBlog, Blog V4, SN: 00000001

Using device 0: Generic RTL2832U OEM
Found Rafael Micro R828D tuner
RTL-SDR Blog V4 Detected
Tuner gain set to automatic.
Tuned to 97571000 Hz.
Oversampling input by: 6x.
Oversampling output by: 1x.
Buffer size: 8.03ms
Exact sample rate is: 1020000.026345 Hz
Sampling at 1020000 S/s.
Output at 170000 Hz.
Playing raw data 'stdin' : Signed 16 bit Little Endian, Rate 32000 Hz, Mono
^CSignal caught, exiting!

User cancel, exiting...
Aborted by signal Interrupt...
pi@raspberrypi:~ $ 
```

Figure 6.8: Tuning to a frequency manually.

In the remaining chapters of the book, you'll discover some of the popular amateur radio programs for the RTL-SDR dongles and learn how to install and use them.

6.8 Testing the RTL-SDR Dongle

You can enter the command **rtl_test** to test your RTL-SDR dongle. Figure 6.9 shows a typical output displayed on Raspberry Pi 5.

```
pi@raspberrypi:~ $ rtl_test
Found 1 device(s):
  0:  RTLSDRBlog, Blog V4, SN: 00000001

Using device 0: Generic RTL2832U OEM
Found Rafael Micro R828D tuner
RTL-SDR Blog V4 Detected
Supported gain values (29): 0.0 0.9 1.4 2.7 3.7 7.7 8.7 12.5 14.4 15.7
 20.7 22.9 25.4 28.0 29.7 32.8 33.8 36.4 37.2 38.6 40.2 42.1 43.4 43.9
 49.6
Sampling at 2048000 S/s.

Info: This tool will continuously read from the device, and report if
samples get lost. If you observe no further output, everything is fine.

Reading samples in async mode...
^CSignal caught, exiting!

User cancel, exiting...
Samples per million lost (minimum): 0
pi@raspberrypi:~ $ 
```

Figure 6.9: Testing the RTL-SDR dongle.

Chapter 7 • A Simple FM Receiver

7.1 Overview

This is a few lines of Python program developed by the authors, which should enable you to tune to a broadcast frequency in the VHF FM band (88–108 MHz) and then listen to the broadcast.

7.2 The Program

The program is based on the RTL_FM and Figure 7.1 shows the program listing (Program: **simpleFM.py**) which consists of only a few lines of code. The program prompts the user to enter the required frequency in kHz and then uses RTL_FM to tune to the frequency entered. Copy the program to your Raspberry Pi 5, plug in your RTL-SDR, and enter the following command to launch the program:

pi@raspberrypi:~ $ **python3 simpleFM.py**

```
#================================================================
#                       TUNE TO FM FREQUENCY
#                       ====================
#
# This program tunes to an FM frequency (WBFM) using the RTL_FM
#
# Author: Dogan Ibrahim (G7SCU)
# File  : simpleFM.py
# Date  : February, 2024
#=================================================================
import os
print("")
print("SimpleFM - Tune to a FM frequency")
print("=================================")
print("")

freq = 1000.0*float(input("Enter the FM frequency (kHz): "))
frequency = str(freq)
rtl = "rtl_fm -M wbfm -f" + frequency + " | aplay -r 32000 -fS16_LE -c 1"
os.system(rtl)
```

Figure 7.1: Program listing.

Figure 7.2 shows an example run of the program where the required frequency was 97.3 MHz (frequency of the LBC radio broadcast in London). Enter **Cntrl+C** to terminate the program.

```
pi@raspberrypi:~ $ python simpleFM.py

SimpleFM - Tune to a FM frequency
==================================

Enter the FM frequency (kHz): 97300
Found 1 device(s):
  0:  Realtek, RTL2838UHIDIR, SN: 00000001

Using device 0: Generic RTL2832U OEM
Found Rafael Micro R820T tuner
Tuner gain set to automatic.
Tuned to 97571000 Hz.
Oversampling input by: 6x.
Oversampling output by: 1x.
Buffer size: 8.03ms
Exact sample rate is: 1020000.026345 Hz
Sampling at 1020000 S/s.
Output at 170000 Hz.
Playing raw data 'stdin' : Signed 16 bit Little Endian, Rate 32000 Hz, Mono
```

Figure 7.2: Example run of the program.

7.2.1 Creating a shell script

An alternative to creating a Python program is to create a shell script and run RTL_FM from this script. The steps are:

- Use the **nano** editor and create a file called **fm** with the contents as follows:

```
!/bin/sh
rtl_fm -M wbfm -f$1 | aplya -r 32000 -fS16_LE -c 1
```

- Make the script executable:

pi@raspberrypi:~ $ **sudo chmod +x fm**

- Run the script file and specify the frequency. For example, if the frequency is 97.3 MHz, enter:

pi@raspberrypi:~ $ **./fm 97.3M**

or,

pi@raspberrypi:~ $ **sh fm 97.3M**

Notice that **$1** in the script corresponds to the entered frequency value.

Chapter 8 • GQRX on the Raspberry Pi 5

8.1 Overview

GQRX is an excellent open source software defined radio receiver (SDR) created by Alexander Csete (OZ9AEC) and powered by the GNU Radio and the Qt graphical toolkit. In this chapter, you'll see how to install the GQRX on your Raspberry Pi 5 and learn how to use it. GQRX offers the following features:

- Change frequency, gain and apply various corrections
- AM, SSB, CW, FM-N and FM-W demodulators
- FM ode for NOAA APT
- Variable band-pass filter
- FFT plot and waterfall
- AGC, squelch and noise blankers
- Record and playback raw baseband data and audio to/from WAV file
- Spectrum analyzer
- Remote control through TCP
- Streaming audio output over UDP
- Discover devices attached to the computer

8.2 Installation on Raspberry Pi 5

The steps to install the latest version of GQRX on your Raspberry Pi 5 are given below:

- sudo apt-get update
- sudo apt-get install -y cmake gnuradio-dev gr-osmosdr qt6-base-dev qt6-svg-dev qt6-wayland libasound2-dev libjack-jackd2-dev portaudio19-dev libpulse-dev
- git clone https://github.com/gqrx-sdr/gqrx.git
- cd gqrx
- mkdir build
- cd build
- cmake ..
- make
- sudo make install
- volk_profile

Press Cntrl+C to terminate **volk_profile**, Notice that the **volk_profile** command is used to optimize the signal processing routines for best performance, and to work around a bug that would otherwise prevent the AM demodulator from working.

8.3 Using the GQRX

Before using the GQRX, make sure that your RTL-SDR dongle, USB audio adapter, and the antenna are all connected as required, and proven to be operational. This is important since the dongle and the audio adapter are checked at the beginning of the program and the program may not recognize them if they are connected after it starts.

The steps are:

- Start your Desktop. If you do not have a directly connected monitor with HDMI cable, then start the **vncserver** and then the **TightVNC Viewer** so that the Desktop is displayed on your PC. If you have a directly connected monitor and keyboard then just continue.

- Start a terminal session on your Desktop and type:

  pi@raspberrypi:~ $ **gqrx**

Figure 8.1 shows the GQRX main screen tuned to 97.3 MHz, which happens to be the broadcast frequency of London LBC radio which is local to the authors. You will see the following menu items at the upper part of the screen:

- File
- Tools
- View
- Help

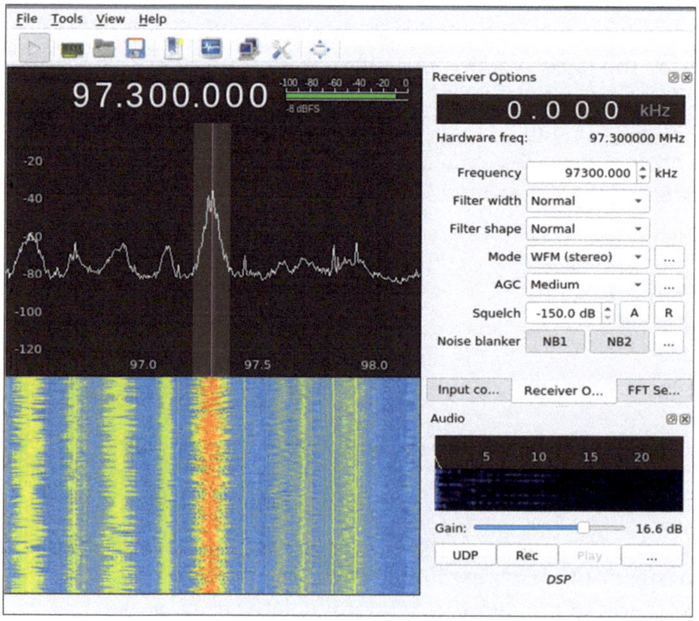

Figure 8.1: GQRX main screen.

By selecting the File menu you can configure the I/O devices, load saved settings, save the current settings, save the waterfall, or quit the program. Click **File → Start DSP** to start receiving radio signals at the selected frequency (or click the grey Start/Stop button under File). The **I/O devices** menu (Figure 8.2) is the first displayed menu. This menu allows you to select items such as the type of device you are using, input rate, audio device, and audio sample rate. In this example, the **Realtek RTL2838UHIDII** is selected as the device (this is the RTL-SDR V3 dongle). The **Input rate** is set to 1800000 but you can experiment with different values. The **Audio output** is set to **Default** (the authors had the UGREEN audio adapter connected to one of the USB ports of the Raspberry Pi 5).

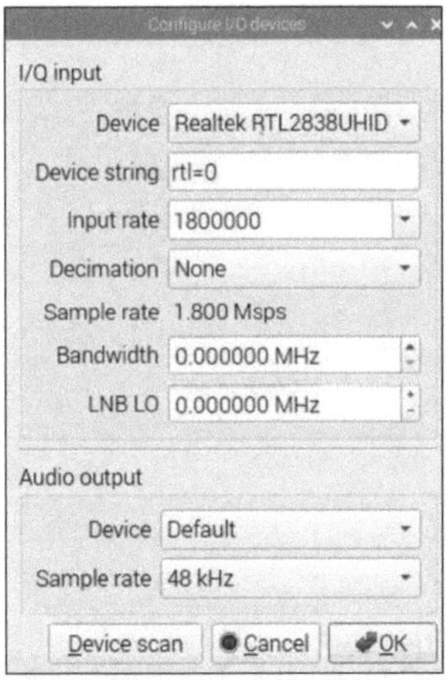

Figure 8.2: The I/O device menu.

Near the bottom right of the screen, you will see 3 tabs with the names: **Input Controls**, **Receiver Options**, and **FFT Settings**.

At the top of the **Input Controls** (Figure 8.3) is the **LNB LO** which is the local oscillator frequency in front of your SDR dongle, and it is set to 0 in most cases.

The **Hardware AGC** should be ticked as it enables automatic gain control.

Swap I/Q is used to swap I and Q channels and it should normally not be ticked.

No limits must not be ticked as it may enable use of the SDR beyond its normal frequencies (you may have to enable it to receive HF frequencies).

DC remove removes the DC bias and it should be ticked.

IQ balance should not be ticked unless there seems to be ghost images in the spectrum.

Freq Correction is used to correct for the drift in the SDR internal oscillator. This is adjusted by tuning the dongle to a very well-known stable frequency (e.g. using a frequency generator) and then adjusting this correction until the displayed frequency matches the frequency of the generator. In most cases, you can leave this set to 0.0. After making a correction, you should save the settings.

Antenna should be kept at **RX** as there is usually no other options with most dongles,

You should tick the **Reset frequency controller digits**.

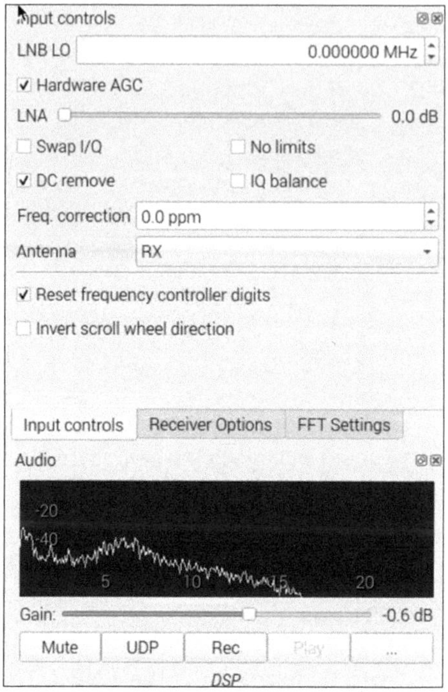

Figure 8.3: Input controls.

At the top of the **Receiver Options** (Figure 8.4) you can offset the selected frequency from the main frequency entered in the left window. This is normally kept at 0.000.

Filter width and **Filter shape** can be selected as required. They are usually set to **Normal**.

Mode is used to select the require modulation mode and it can be:

- **Demod Off** turns off signal processing (the real time spectrum can still be viewed).
- **Raw I/Q** is selected to pass raw I/Q data without any demodulation.
- **Narrow FM** used to select narrow FM.
- **WFM** (mono or stereo) is used to select broadcasting stations in the FM band
- **USB** or **LSB** are the upper and lower side bands, selected for side band amplitude modulation.
- **CW-L** or **CW-U** are selected for Morse code modulation.

AGC should be selected as required.

Squelch value can be set manually, or you can wait until there is silence in the band and then click on the button **A** to its right to automatically adjust the squelch level to the current signal or noise level. Clicking **R** rsets the squelch level to its default value.

Noise blanker **NB1** and **NB2** attenuate noise.

The **Audio** section at the bottom of the **Receiver Options** controls the audio signal level by using the slider. Set the audio level to several dB. The audio signal can be recorded by clicking the **Recording** tab. The **UDP** tab allows you to stream the raw audio over UDP connection.

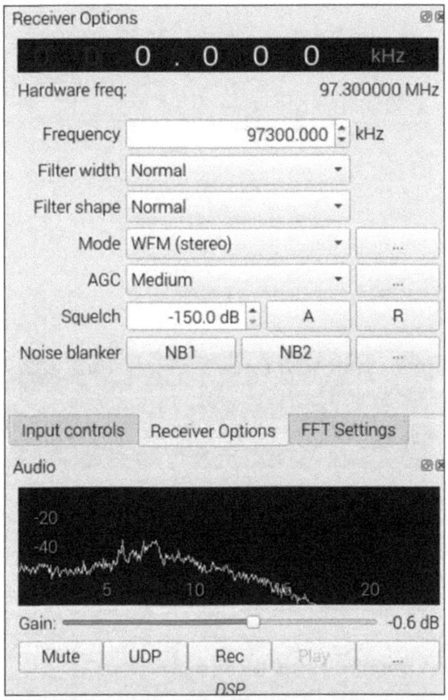

Figure 8.4: Receiver options.

The **FFT Settings** window is shown in Figure 8.5. **FFT Size** is the number of points used in the FFT calculation and it determines the resolution of the plot. For example, if the bandwidth is 200 kHz and FFT size is 8192, then the FFT resolution is 200000 / 8192 = 24 Hz / FFT which is related to the screen pixels

FFT rate is the frames per second and it can be selected to up to 60 fps.

Window is the type of windowing used in the DSP and by default it is set to Hann.

FFT Averaging is the averaging gain used to reduce the noise level. The averaging affects the FFT plot and not the **waterfall**.

The **Peak Detect** button will show little circles at the peaks of the spectrum display (Figure 8.6). **Band Plan** will show the band details (Figure 8.7).

The **WF dB** sets the **waterfall** gain in dB.

Freq zoom sets the frequency axis scale

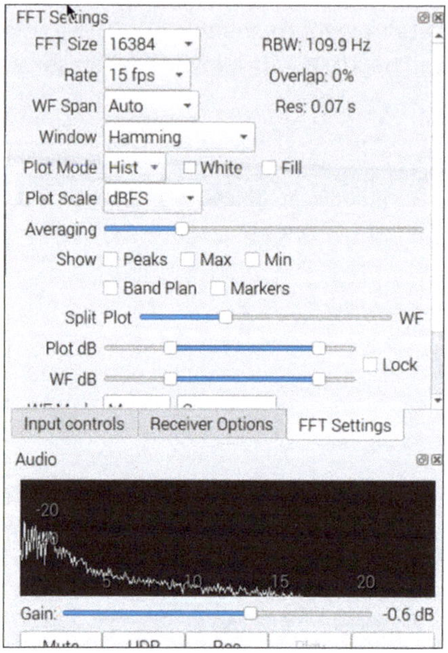

Figure 8.5: FFT settings options.

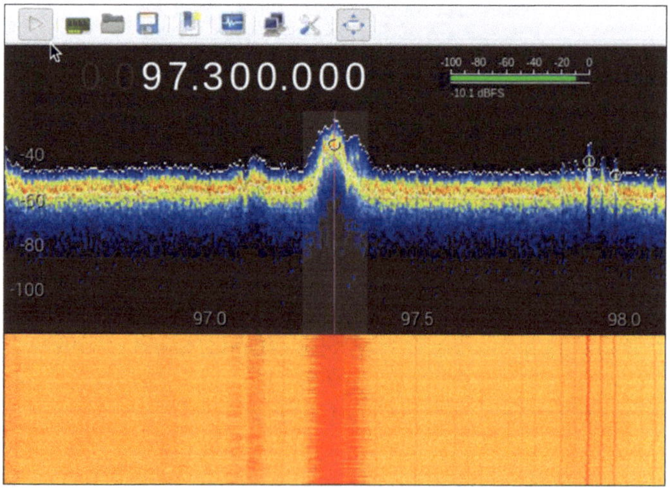

Figure 8.6: Peak Detect selected.

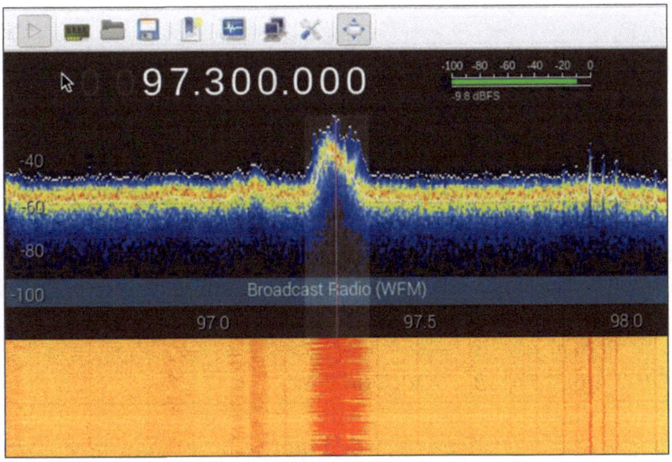

Figure 8.7: Band Plan selected.

8.3.1 The audio frame

In the lower right window appears either the spectrum and controls of the Audio signal received (Figure 8.8) or the FFT Settings depending on which tab you clicked on. The Audio spectrum frame shows the spectrum of the demodulated signal. You can change the scale of the spectrum by click and drag the mouse pointer of the frequency scale, and move it right-left with the mouse wheel. The **slider** below the audio spectrum control the output level to the speaker/headphone,

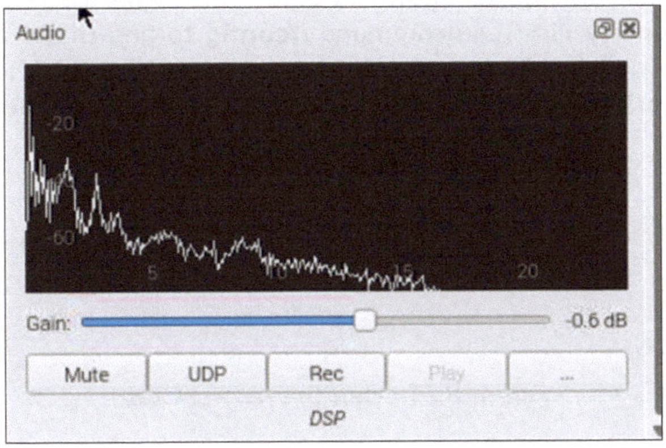

Figure 8.8 Audio frame

Clicking the **Mute** button stops output to the speaker/headphone. **Rec** (recording) button is used to record the received audio. The audio file name format is:

gqrx_yyyymmdd_hhmmss_frequency.wav

where,

yyyy is the year
mm is the month
dd is the day
hhmmss the start hour, minutes,seconds
frequency is the hardware in Hz

For example, click the **Rec** button for a short while the reception is in progress. The filename will be displayed under the Rec button. Click the button again to stop the recording. A file will be created in your **/home/pi** directory. As an example, the following file was created in author's test:

gqrx_20240205_151413_97300000.wav

in the above example, the audio was recorded on the 5th of February 2024 (20240205) at 15:14:13 hours and the receiver was set to 97.3 MHz.

You can click the **Play** button to playback a previously recorded file.

8.3.2 Streaming audio to your PC

GQRX has the option of streaming the audio data over the network using the UDP protocol. This is, for example, useful if you are having problems outputting audio directly to your speakers with the GQRX. Before streaming the data, we have to configure the GQRX network settings. Open the network window by clicking the audio settings in the Audio window. Click the network tab as shown in Figure 8.9, where 192.168.1.202 is the IP address of the Raspberry Pi 5 (use command **ifconfig** to find the IP address it if you do know what it is). Click the bottom right-hand tab (**Audio settings**) under the Audio tab and then the **Network** tab to start outputting the audio from the GQRX.

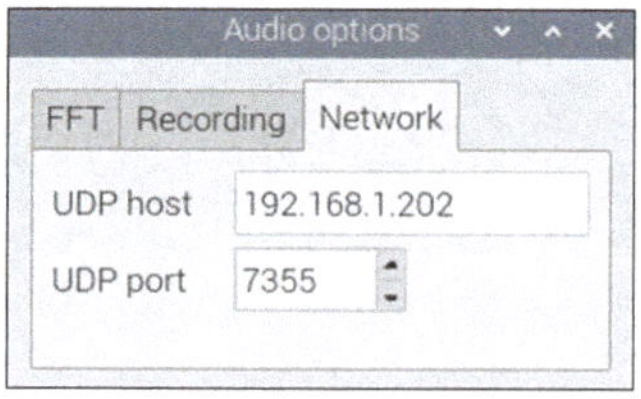

Figure 8.9: Open the network tab

Then enter the following command (see Figure 8.10) at the command line to stream the audio on your Raspberry Pi 5 speakers. Enter **Cntrl+C** to terminate the streaming:

pi@raspberrypi:~ $ **nc -l -u 7355 | aplay -r 48000 -f S16_LE -t raw -c 1**

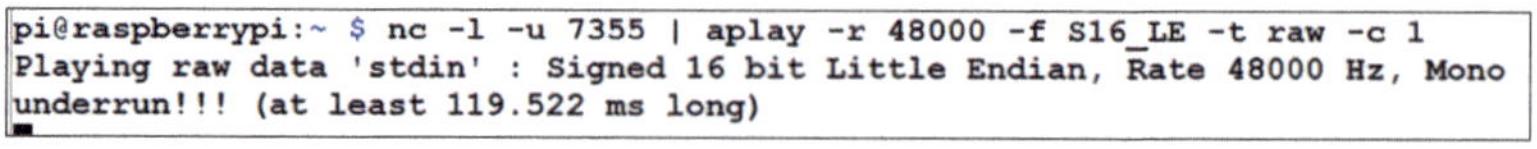

Figure 8.10: Start the audio streaming.

You could get a better audio quality by sending the output to the **VLC** media play program as shown below (Figure 8.11). Enter **Cntrl+C** to terminate the streaming:

pi@raspberrypi:~ $ **vlc --demux=rawaud --rawaud-channels=1 --rawaud-samplerate=48000 udp://@:7355**

```
pi@raspberrypi:~ $ vlc --demux=rawaud --rawaud-channels=1 --rawaud-samplerate=48
000 udp://@:7355
VLC media player 3.0.11 Vetinari (revision 3.0.11-0-gdc0c5ced72)
```

Figure 8.11: Sending the output to VLC.

Chapter 9 • SDR++

9.1 Overview

SDR++ (or "SDR PlusPlus") is an open source, cross-platform, software-defined radio program that works with the RTL-SDR as well as many other SDR receivers. It is a popular and easy to use program, which can be installed and operated on many platforms.

SDR++ has the following features (*SDR++ User Guide*):

- Wide hardware support (e.g. Raspberry Pi, Windows, MAC etc.).
- Full waterfall.
- Listening to more than one frequency at the same time within the receive bandwidth.
- Remote operation of the SDR using the built in server.
- Range scanner.

In this chapter, you explore the installation and practical use of the SDR++ on the Raspberry Pi 5.

9.2 Installing SDR++ on the Raspberry Pi 5

The steps are:

- sudo apt update
- sudo apt install -y build-essential cmake git libfftw3-dev libglfw3-dev libglew-dev libvolk2-dev libsoapysdr-dev libairspyhf-dev libairspy-dev \
- libiio-dev libad9361-dev librtaudio-dev libhackrf-dev librtlsdr-dev libbladerf-dev liblimesuite-dev p7zip-full wget
- git clone https://github.com/AlexandreRouma/SDRPlusPlus
- cd SDRPlusPlus
- sudo mkdir -p build
- cd build
- sudo mkdir -p CMakeFiles
- sudo cmake .. -DOPT_BUILD_RTL_SDR_SOURCE=ON
- sudo make
- sudo make install

The **git clone** command requires your GitHub username and token number. If you don't have a github account you should create one using your PC after giving a username and password. The steps to get a token number are as follows:

- log in into your GitHub account.

- go to **Settings → Developer Settings → Personal Access Token → Tokens (classic)**

- **Generate token**.

- copy the generated token and use it where the password is requested. The token will be something like: ghp_sFhFsSHh0ytzMDRLjmks4yy76gf59zgthdvfsrta

9.3 Using the SDR++

The SDR++ program requires a directly connected monitor (using a HDMI cable) to work. You cannot connect using a PC with vncserver and TightVNCViewer in Desktop mode.

To start the SDR++ follow the steps:

- Connect your Raspberry Pi 5 to a monitor using a HDMI cable.
- Connect the RTL-SDR dongle, antenna, audio adapter, keyboard, and mouse.
- Boot your Raspberry Pi 5. If it starts in console mode enter the following command to start the Desktop mode:

  pi@raspberrypi:~ $ **startx**

- Open a terminal session in Desktop mode and enter the following command to start the SDR++:

  pi@raspberrypi:~ $ **sdrpp**

Figure 9.1 shows the SDR++ startup screen.

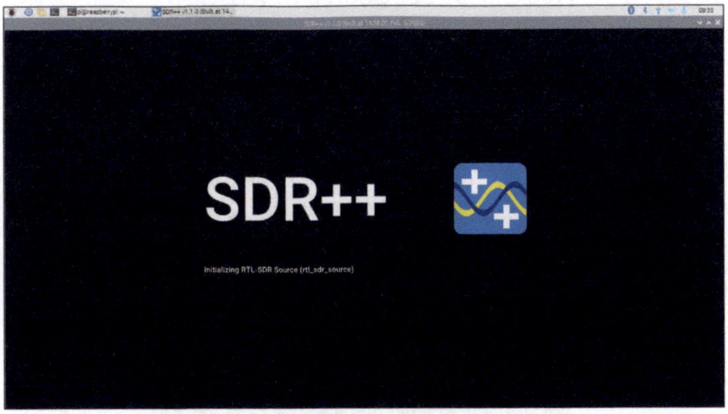

Figure 9.1: SDR++ startup screen.

9.3.1 Quick startup example

Perhaps the easiest way to learn to use the SDR++ is to look at an example. In this example, we will tune to 97.3 MHz which is the frequency of radio station "LBC" in London. The steps are as follows:

- Click at the bottom or top of the frequency numbers and set it to 97.300.000 as shown in Figure 9.2.

Figure 9.2: Set the required frequency.

- Click **Source** and set the SDR receiver type to **RTL-SDR** (Figure 9.3).

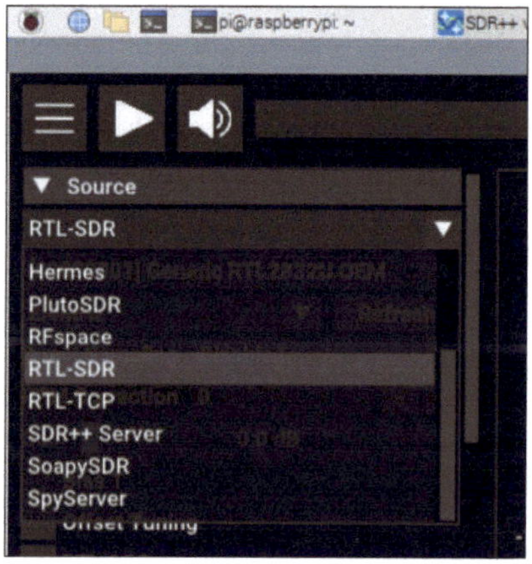

Figure 9.3: Set Source to RTL-SDR.

- Click **Source** and click to enable **Tuner AGC** (Figure 9.4).

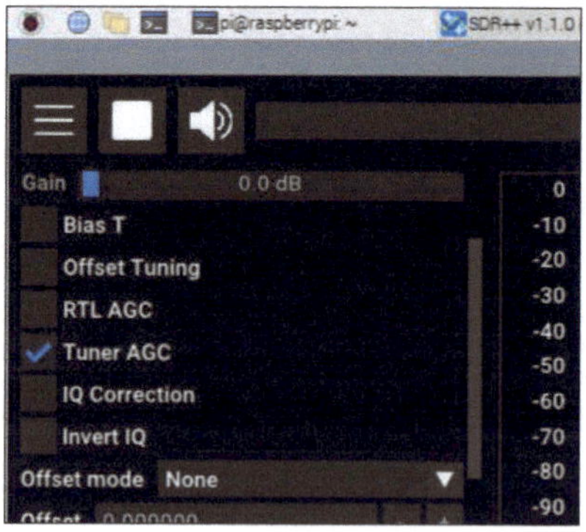

Figure 9.4: Enable Tuner AGC.

- Expand **Radio** and select **WFM** (Figure 9.5).

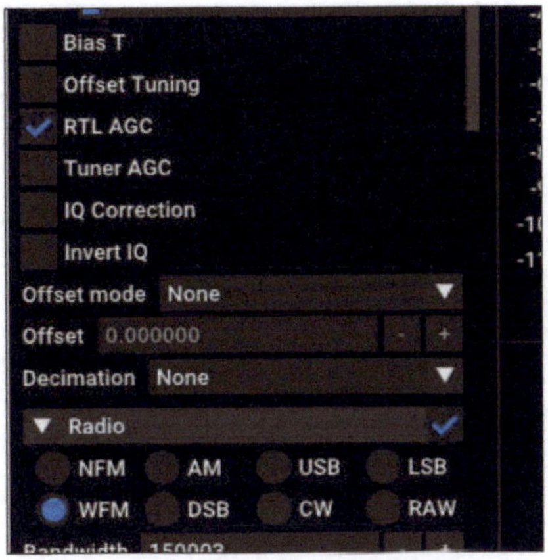

Figure 9.5: Enable WFM.

- If required, select **Stereo**, **Low Pass** (filter) etc. (Figure 9.6).

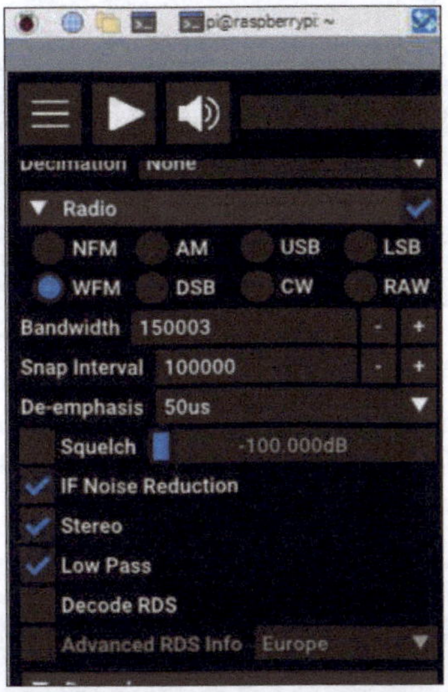

Figure 9.6: Select Stereo, etc.

- Click **Display** and enable **Show Waterfall** (Figure 9.7).

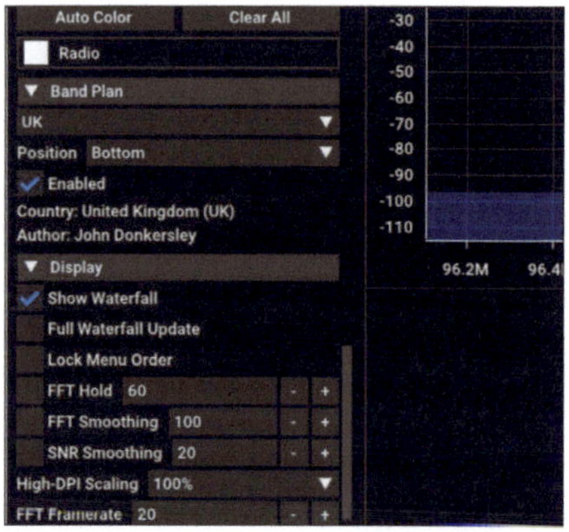

Figure 9.7: Enable Waterfall.

- Click **Audio** and select **Audio Adapter** (Figure 9.8). Note that if you select the **Built-in Audio** then the sound will be output from your monitor.

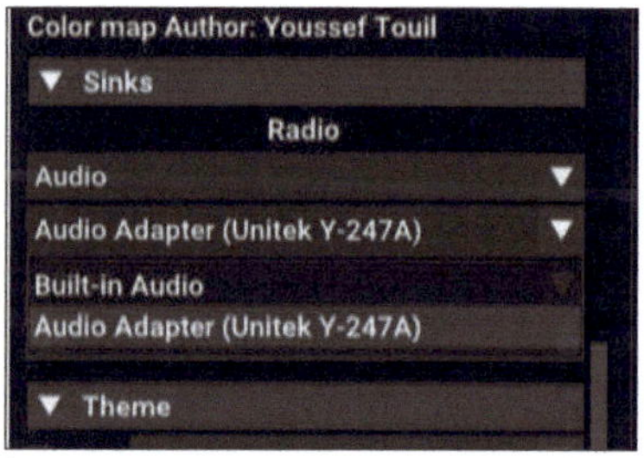

Figure 9.8: Select your audio adapter.

- Finally, click the right arrow button at the top left of the screen to start the reception. You should hear the radio on your speakers. Figure 9.9 shows the reception with the waterfall.

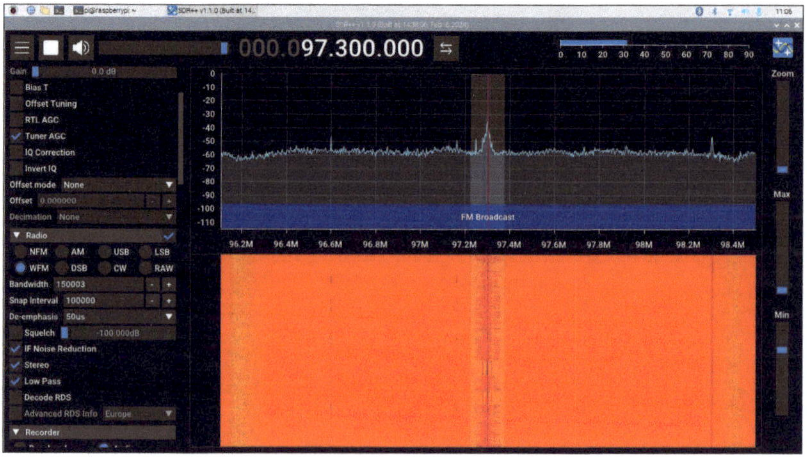

Figure 9.9: Reception using the "waterfall" display.

In the remaining parts of this chapter, we will look at other options of SDR++ in some detail. Notice that if an option is greyed out, it cannot be changed.

9.3.2 Graphical outputs

Figure 9.10 shows the top part of the display where various graphical items are identified.

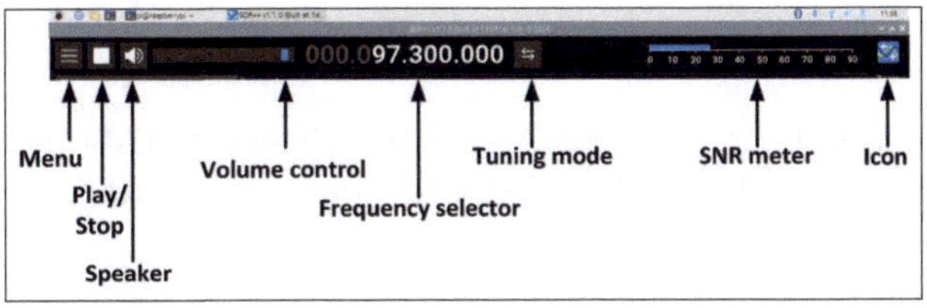

Figure 9.10: Top part of the display.

SNR Meter

The SNR meter is not calibrated, it just shows a measure of the relative strength of signals above the background noise.

Frequency Tuning

The received can be tuned in several ways.

- Place the mouse cursor over a digit that you want to change and change it using the scroll wheel.
- Click at the top or bottom of a digit to increase or decrease the frequency, respectively.
- Place the mouse pointer over the largest digit you want to change and type in the complete frequency you want to set to.
- Drag the frequency scale at the bottom of the FFT spectrum display.

Tuning Mode

The tuning can either be **Centre tuning** or **Normal tuning**. With the Centre tuning, when you click or tap on the waterfall (or spectrum) to tune, the selected frequency snaps to the center of the display, which may be the preferable option. In Normal tuning mode you can tap or tune anywhere on the spectrum display and they stay where they are.

The icon

The icon at the top right of the display shows the credits and the build version of the SDR++.

The spectrum display

You can see several receptions on the display and select which one you want to investigate. The vertical axis shows the signal strength while the horizontal axis shows the frequency. You can see the strength of each signal through the depth of the colour and the width of the transmission. You can tap on a peak to select that frequency. You can move the sliding bar at the right hand side of the display (marked **Max**) up or down to change the range of the frequency scale. The spectrum display is also useful to match a manual antenna tuner for maximum signal strength. You should adjust the antenna tuner until you get the maximum peaks on the spectrum.

9.3.3 Source options

AGC

This is the Automatic-Gain-Control. The AGC automatically adjusts the gain so that weak signals become stronger and the gain of strong signals is reduced. This helps to have a comfortable gain.

Sample Rate

This shows how much the frequency spectrum your SDR can show for the selected frequency. This depends on the type of SDR device used. Most RTL-SDR dongles can show signals across to up to about 2 MHz.

Offset Tuning (for up-down converters)

Up-conversion and down-conversion are done using additional external hardware connected to your SDR dongle. For example, if your SDR receiver covers VHF 30 – 300 MHz and you want to listen to say lower than 30 MHz frequencies, then you can use an up converter to raise the frequencies so that your SDR can receive them. You can click the offset mode to see the true frequency of the signal received. Do not click the offset mode if you are not using a converter.

IQ correction

Adjust this value if you can see effects such as spikes on the frequency display, or imbalances that may affect the audio. You should adjust the value until the unwanted spikes or imbalances disappear.

Decimation

This should normally be at zero. It reduces the number of times a sample is taken of the radio signals received. For example, 2 means that 1 in every 2 samples will be taken. Decimation can increase the dynamic range of the receiver. You may like to try different settings.

Gain Control

The LNA, or the gain control, is used for gain control. If you have selected AGC then you will not be able to adjust the gain manually. By adjusting the gain manually you can boost the received signal and this is shown in decibels (dB). The gain is increased by dragging the blue bar to the right and reduced by dragging it to the left. When you select a different frequency, you may have to adjust the gain manually again unless the AGC is selected.

Direct Sampling

You can select lower frequencies by using direct sampling. Although this works, the performance may become poor (poorer performance than using an up converter). In order to receive frequencies in the range 500 kHz – 24 MHz, direct sampling can be used, but when this is selected you will not be able to use VHF and UHF band signals.

PPM Correction

The oscillator of an RTL-SDR device may not be stable, especially this is true for inexpensive devices. More expensive RTL-SDR devices use temperature controlled crystal oscillators for accuracy. You can use the PPM correction to calibrate your receiver if it happens not to be accurate. The correction is in parts per million. The way to calibrate your RTL-SDR in SDR++ is to use the **rtl_test** utility.

9.3.4 Display options

The display options control the waterfall and the FFT features of the program.

Show Waterfall

Used to turn the waterfall display ON/OFF.

Fast FFT

Use this to change the rendering speed, e.g., show less detail in the waterfall and spectrum displays.

Full Waterfall Update

Enable it to update the history of the waterfall when zooming or min/max changes

FFT Hold

Use this to display the maximum signal level that has been reached. The trace duration can be adjusted.

Lock Menu Order

Enable this so that the menu section cannot be moved.

High-DPI Scaling

Using this option you can control the size of the controls, scroll bars, and text (e.g., if you have a high resolution monitor).

FFT Frame Rate

The default of 20 should be ok. It can be set to produce more detailed FFT/Spectrum display at the expense of frequency resolution.

FFT size

Use this option to increase or decrease the waterfall/spectrum display. Higher will be better as it can show weaker signals, but it will require more processing power.

FFT Window

This sets the algorithm used for the FFT.

9.3.5 Radio module

These options set the modulation type. The options are shown in Table 9.1 (taken from the *SDR++ User Guide – For SDR++ up to version 1.1*, December 2022.

Mode	Name	Use	Default Filter
NFM	Narrow band Frequency Modulation	From 27 MHz upward for Citizens Band, most business radio and on amateur radio bands. Voice, and V/UHF data transmissions such as DMR, NOAA APT and POCSAG/Flex are modulated in NFM	12.5 khz
WFM	Wideband Frequency Modulation	Broadcast band 87-108 MHz. You can also choose the de-emphasis used, and switch stereo reception on and off.	150 kHz
AM	Amplitude Modulation	Long, medium and short wave broadcast stations, Citizens Band, Airband.	10 kHz
DSB	Double Sideband	Used for dual sideband suppressed AM carrier transmissions.	4.6 kHz
USB	Upper sideband	Amateur bands above 10MHz and on 60m for voice, and for most data modes on all bands; most HF utility stations (air, marine, military)	2.8 kHz
LSB	Lower sideband	Mostly used for voice transmissions on Amateur bands below 10 MHz and for rtty (radioteletype)	2.8 kHz
CW	Continuous Wave	Morse code transmissions. Basically LSB with a narrow filter	200 Hz
RAW		This outputs the IF of the VFO as audio. Left channel is I branch, right is Q branch	

Table 9.1: Modulation types (source: https://sdrpp.org).

Squelch

The squelch stops the audio if the signal does nor exceed a chosen level. Squelch is used to mute background noise and to allow wanted signals to pass. The squelch can be set by ticking the box to activate it and then dragging the marker.

Snap Interval

Use this option to set the frequency step during tuning.

Stereo

Use this option to toggle between stereo and mono (WFM only).

Low Pass

Use this option to apply filter to the received audio signal. The filter removes frequencies exceeding a modulation index of 1.

9.3.6 Frequency Manager

This can be used to save and display your favourite frequencies in bookmarks (Figure 9.11). Bookmarked frequencies can be displayed in the spectrum. You can create lists in the frequency manager (there is by default a General list). For more details refer to the *SDR++ User Guide*.

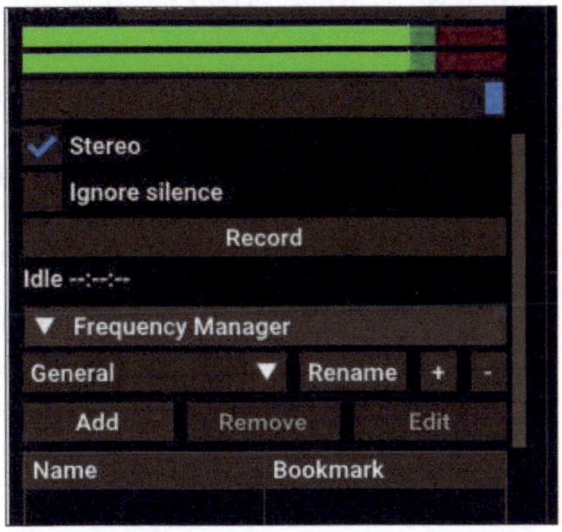

Figure 9.11: Frequency manager.

9.3.7 Recorder

The Recorder option allows the audio you are tuned to be recorded, or the baseband to be recorded (The baseband is the whole of the raw IQ data from the bandwidth the SDR is processing). By default the recorder saves the file in a sub-folder **recordings** of the directory where SDR++ is installed. In Raspberry Pi 5 this is in **~/.config/sdrpp/recordings**. There is a signal strength (S-) meter to enable you to adjust the output so that only the strongest peaks go into the red.

Press **Record** to start recording, the button will change into a **Stop** button, which can be clicked to stop recording. The recorded files are stored in the following format:

audio_frequency_hh-mm-ss_dd-mm-yyyy/wav

For example, if the receiver is tuned to 97.3 MHz and the recording was done at 14:21:05 on 7th of February 2014, the file will heve the format:

audio_97300000Hz_14-21-05_07-02-2024.wav

Chapter 10 • CubicSDR

10.1 Introduction to CubicSDR

CubicSDR is a cross-platform, software-defined radio receiver that allows you to tune across the radio spectrum within the bands of the RTL-SDR dongle. CubicSDR supports many SDR hardware devices such as SDRPlay, HackRF, BladeRF, AirSpy, Red Pitaya, and of course the RTL-SDR dongle.

The **CubicSDR** is probably one of the best SDR software for the beginners, and yet it includes almost everything that a user may want. **CubicSDR** supports most RTL-SDRs.

The **CubicSDR** runs nicely on the Raspberry Pi 5 Desktop monitor (it does not run on a PC logged in via the vncserver and TightVNC Viewer). The installation steps are simple, as follows:

- Start the Desktop on your Raspberry Pi 5.
- Click **Preferences → Add/Remove Software**.
- Enter **CubicSDR** in the search box and press **Enter**.
- Click to select the software and click **Apply.**
- Enter your Raspberry Pi 5 password and wait until the installation is finished.
- To run the software, start a terminal session by clicking the Terminal menu in Desktop and enter **CubicSDR** and press Enter. Alternatively, click **Other** from the **Applications menu** on Desktop and click on **CubicSDR** icon.
- Select **SDR Devices** and then double click on the **RTL...** to start the receiver.

10.2 Quick Startup

In this example, we tune to 97.3 MHz which is the frequency of the London LBC radio station. The steps are:

Start the CubicSDR as described above and double click on **RTL...** in **SDR Devices** menu.

Click anywhere on the waterfall to start the receiver. Note that by default the sound will be output on your monitor speakers (through the HDMI cable). You can change this to your audio adapter if you wish by selecting **Audio Out** from the FM settings at the top left of the screen next to **FM**.

Select the mode, e.g., FM, enter the required frequency e.g., 97.3 MHz (entered as 93700000). You can click at the bottom or top of a frequency number in tuning menu to set the required frequency. Alternatively, click on the top frequency digit and then enter the required frequency through the keyboard.

Figure 10.1 shows the **CubicSDR** running at 97.3 MHz FM. Various parts of the screen are also identified in this figure (further details on the **CubicSDR** can be obtained from the web site: https://cubicsdr.readthedocs.io/en/latest/application-window.html).

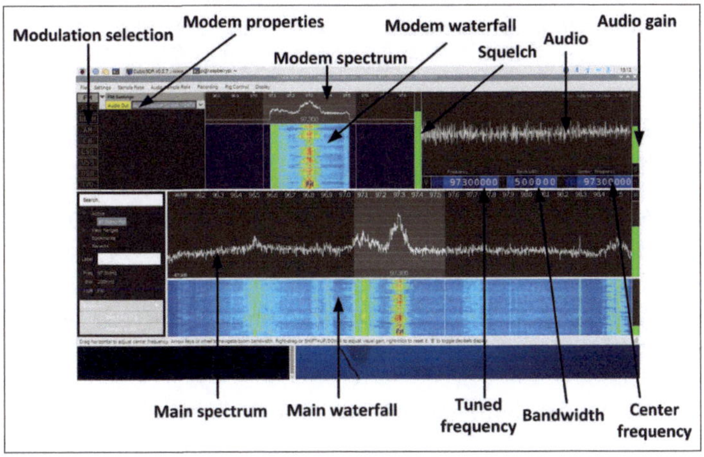

Figure 10.1: CubicSDR running at 97.3 MHz.

The main spectrum and waterfall displays can be zoomed in or out using the arrow keys or the mouse wheel. **CubicSDR** uses a fixed-resolution FFT. The Visual Gain can be adjusted by right-clicking and dragging on the Main Spectrum or the Waterfall.

The center frequency is set by dragging left or right on the main spectrum, or by using the left and right arrow keys, or pointing the mouse on a number and using the mouse wheel to change the number.

The modulation can be selected as: AM, FM, FMS (stereo), NBFM, LSB, USB, DSB, or I/Q Raw.

The squelch can be set by dragging the slider at the right side of the Modem Waterfall. Right clicking the squelch will set it just above the current signal level.

The Waterfall speed can be adjusted from 1 to 1024 lines per second by dragging the meter to the right of the main waterfall (or by using the mouse wheel).

Most numeric controls can be entered directly using the keyboard. Just hover over the desired value and press SPACE to open the input dialog, or just start typing the numbers.

When entering the frequency values, if the value is greater than 3000 then Hz will be assumed automatically. Direct input accepts suffixes such as Hz, MHz, or GHz.

The modulation selection has the following options (see the *CubicSDR Documentation* for more details):

- **AM**: Amplitude — AM with carrier signal, Default 6 kHz, Min. 500 Hz, Max. 500 kHz.
- **FM**: Frequency — Default 20 kHz bandwidth, Min. 500 Hz, Max. 500 kHz, Mono.
- **FMS**: Stereo Bandwidth — Default 200 kHz, Min. 100 kHz, Max. 500 kHz, Stereo (multiplex).
- **NBFM**: Narrow-Band Frequency Modulation — Default 12.5 kHz, Min. 500 Hz, Max. 500 kHz, Mono.
- **LSB**: Lower Side Band — Lower sideband of AM (no carrier), Default 2.7 kHz, Min. 250 Hz, Max. 250 kHz.
- **USB**: Upper Side Band — Upper sideband of AM (no carrier), Default 2.7 kHz, Min. 250 Hz, Max. 250 kHz.
- **DSB**: Dual Side Band — Same as AM but without carrier signal, Default 5.4 kHz, Min. 500 Hz, Max. 500 kHz.
- **I/Q**: Raw I/Q Pass-Thru (no modulation) — Raw I/Q samples that would normally go to a modem are passed through to the sound card for use elsewhere. Bandwidth is fixed to the selected sound card output frequency and will change along with it.

Chapter 11 • RTL-SDR Server

11.1 Overview

The concept of RTL-SDR is highly interesting as it allows an RTL-SDR dongle to be plugged into, say, a Raspberry Pi 5 computer, and then the received data is sent over the network using the TCP protocol to a SDR software running on another computer. RTL-SDR server could be used, for example, if you have your Raspberry Pi 5 together with the antenna at a remote location, and you wish to access the RTL-SDR remotely, say, out in the garden.

Figure 11.1 shows the concept of an RTL-SDR server using a Raspberry Pi 5 and a PC. In this image, an RTL-SDR dongle is attached to a Raspberry Pi 5 computer together with a suitable antenna. The PC runs a software defined radio program — here, the popular Windows-based **SDR Sharp** program is used.

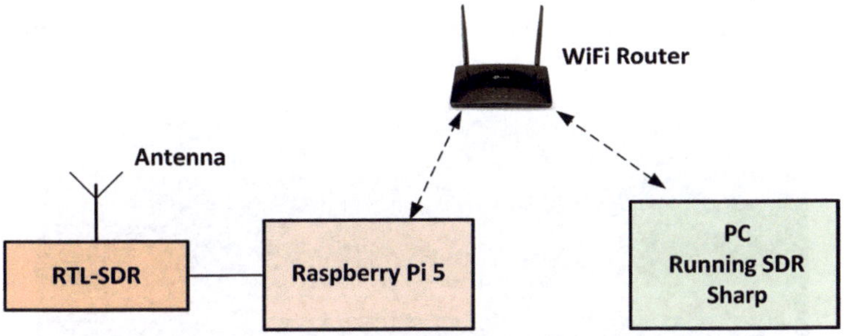

Figure 11.1: RTL-SDR server in action.

The operation of the system is as follows:

- Obtain the IP address of your Raspberry Pi 5, for example, using command **hostname -I** as shown in Figure 11.2. In this example, the IP address was 192.168.1.251.

```
pi@raspberrypi:~ $ hostname -I
192.168.1.251 2a00:23c7:86a3:2501:ac4a:9eca:93e4:a382
pi@raspberrypi:~ $
```

Figure 11.2: Raspberry Pi 5 IP address.

- A command is entered on the Raspberry Pi 5 to start the server process.

- The SDR program on the remote computer (PC) is configured to use the TCP protocol.

The actual steps below describe the operation using commands:

- Plug the dongle into one of the USB ports of your Raspberry Pi.

- Enter the following command with your own IP address:

pi@raspberrypi:~ $ **rtl_tcp –a 192.168.1.251**

where 192.168.1.251 is the IP address of author's Raspberry Pi 5, and 1234 is the port address. You should see a message as in Figure 11.3.

```
pi@raspberrypi:~ $ rtl_tcp -a 192.168.1.251
Found 1 device(s):
  0:  RTLSDRBlog, Blog V4, SN: 00000001

Using device 0: Generic RTL2832U OEM
Found Rafael Micro R828D tuner
RTL-SDR Blog V4 Detected
Tuned to 100000000 Hz.
listening...
Use the device argument 'rtl_tcp=192.168.1.251:1234' in OsmoSDR (gr-osmosdr) so
rce
to receive samples in GRC and control rtl_tcp parameters (frequency, gain, ...)
```

Figure 11.3: Start the server.

- Start **SDR Sharp** on your laptop (assuming you have installed the **SDR Sharp** on your laptop). Figure 11.4 shows the startup screen.

Figure 11.4: SDR Sharp startup screen.

- Select **RTL-SDR (TCP)** under menu **Source** on your **SDR display**(Figure 11.5))

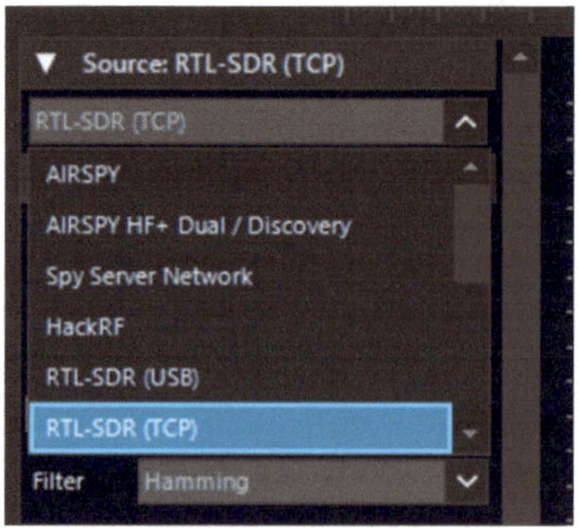

Figure 11.5: Select RTL-SDR (TCP).

- Click the **Configure Source** (Cog wheel icon) and enter the IP address of your Raspberry Pi 5 in the **Host** field, set the port to 1234, and click the Tuner AGC (Figure 11.6).

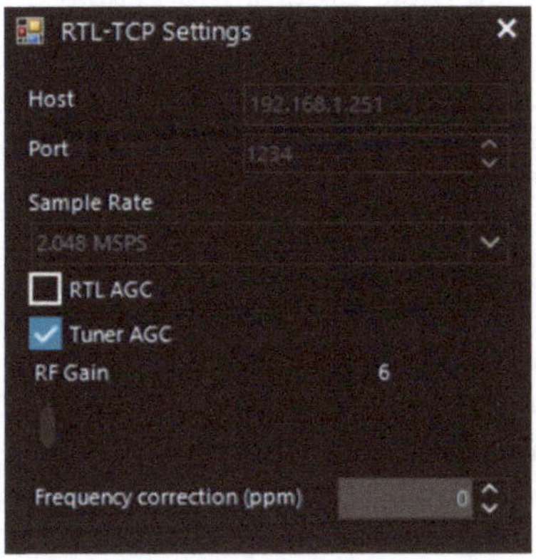

Figure 11.6: Enter the IP address.

- Tune the **SDR Sharp** to the frequency you wish to listen to (e.g., set it to a local FM broadcast frequency for easy checking). You should hear the reception on your PC speakers. Figure 11.7 shows the SDR Sharp display after tuning to the local radio at 97.3 MHz.

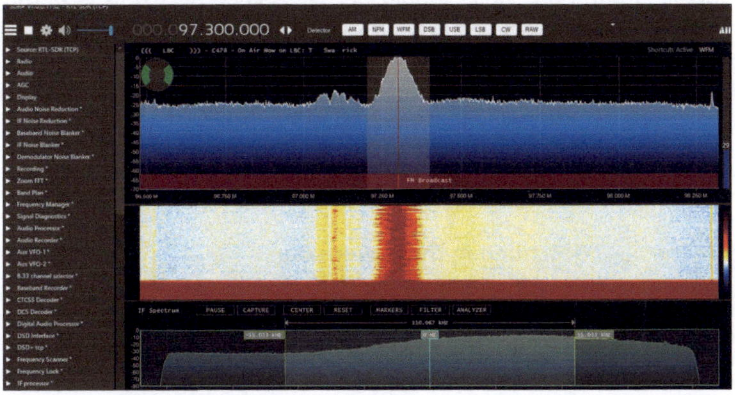

Figure 11.7: SDR Sharp after tuning to 97.3 MHz (example frequency).

Critically evaluated, you may find that the audio quality is not as good as expected. A better response can be obtained if your Raspberry Pi 5 is connected to your network directly (i.e., through an Ethernet cable) rather than over air by Wi-Fi.

Chapter 12 • Dump1090

12.1 Overview

Dump1090 is a software defined radio program, originally written by Salvatore Sanfilippo in 2012 and is used to get real-time air traffic data from aircraft. The program accesses the ADS-B data using a RTL-SDR dongle. The program is available on several platforms such as Windows, Linux etc. The original software was developed by several people and it is now available for use by the Plane Plotter community as well.

12.2 Dump1090 Essential Features

- HTTP support to display the detected aircrafts on Google Maps.
- Ability to decode DF11, DF17 messages.
- Ability to decode DF formats like DF0, DF5, DF16, DF20 and DF21.
- Interactive commands to display detected aircraft.
- CPR coordinates decoding and track calculation from velocity.
- Decoding of weak messages and improved range.
- Ability to decode raw IQ samples from file.
- Plane Plotter interface.
- And more.

A dedicated antenna is recommended for 1.090 GHz operation. The antenna should be mounted as high as possible with a clear view all round (e.g., not in an attic). Additionally, good quality low-loss antenna cable should be used since at 1.090 GHz the cable losses are significant. If you have a decent antenna then you should be able to pick up signals from aircraft very far from your position. With the simple *wine-cork antenna* given in the following website it was reported to pick up signals from over 200 kms away from the local position:

http://antirez.com/news/46

Other good antennas are described in these websites:

http://balarad.net
http://gnuradio.org/data/grcon11/06-foster-adsb.pdf
http://modesbeast.com/pix/adsb-ant-drawing.gif

By default, Dump1090 tries to fix single bit errors using the checksum method.

12.3 Installing Dump1090 on the Raspberry Pi 5

The steps to install the Dump1090 are:

- git clone https://github.com/antirez/dump1090
- cd dump1090
- make

12.4 Launching the Dump1090 Software

Enter the following commands:

pi@raspberrypi:~ $ **cd dump1090**
pi@raspberrypi:~/dump1090 $ **sudo ./dump1090 --interactive**

As shown in Figure 12.1, you should see a dynamic list of the local aircraft with their flight numbers, altitude, speed, latitude, longitude, track, messages, and time observed. The list is updated constantly as more aircraft are detected by the program.

```
pi@raspberrypi: ~/dump1090
Hex     Flight   Altitude  Speed  Lat      Lon       Track  Messages Seen
--------------------------------------------------------------------------------
400b83           5400      199    51.560   -0.071    167    41       0 sec
c0583a           38000     0      0.000    0.000     0      15       7 sec
4cad10  RYR9VJ   30000     392    51.642   0.259     263    186      0 sec
4952c1           3950      141    51.465   -0.199    271    33       9 sec
40658a           2400      113    0.000    0.000     274    11       14 sec
4cc510           24050     400    51.726   -0.176    324    204      0 sec
40104f  EXS6KU   34000     419    51.573   0.319     325    209      1 sec
3c6509  DLH410   32275     434    51.627   0.076     275    284      0 sec
4070eb  EXS4CS   39725     445    51.848   -0.355    349    397      0 sec
4cadfc  RYR57SM  35000     449    51.620   -0.675    351    131      0 sec
49328b           5700      236    51.502   -0.308    8      103      13 sec
400a25           4725      139    51.465   -0.130    273    372      0 sec
406222  CFE7LM   3950      299    51.585   0.089     74     186      4 sec
44a88c  JAF351   35975     453    51.615   -0.121    276    424      1 sec
```

Figure 12.1: List of aircraft detected.

You can display the aircraft on a Google Map after entering the following commands:

pi@raspberrypi:~/dump1090 $ **sudo ./dump1090 --interactive --net**

Make a note of the IP address of your Raspberry Pi 5, and then enter the following command at your Web Browser (e.g., on a PC), where 192.168.1.251 is the IP address of the author:

http://192.168.1.251:8080

Figure 12.2 shows a map of London with the aircraft displayed in real time (it might be hard to see the aircraft in this small map). Click on a plane to display information about it.

Figure 12.2: Aircraft display on the map.

Dump1090 can be called with many other command options to set the gain, frequency, etc. A list of valid commands is obtained by typing:

```
./dump1090 --help
```

Some important options are described below:

• --gain <dB>	set gain (default is max. gain, –100 for auto gain)
• --enable agc	enable automatic gain control (default: off)
• --freq <Hz>	set frequency (default: 1090 MHz)
• --ifile <filename>	read data from file
• --interactive	interactive mode refreshing data every second
• --interactive-ttl <sec>	remove from list if idle (default: 60 seconds)
• --raw	show only message hex values
• --net	enable networking
• --net-only	enable just networking
• --net-http-port <port>	HTTP server port (default: 8080)
• --no-fix	disable single bit error correction
• --metric	use metric units (meters, km/h etc.)

Chapter 13 • FLDIGI

13.1 Overview

FLDIGI is a popular Digital Mode data modem program by David Freese, W1HKJ. It operates in conjunction with a conventional HF SSB transceiver, using audio frequency signals over an audio adapter. Fldigi includes most of the popular digital modes including MFSK16, PSK31, Contestia, DominoEX, CW, FSQ, RTTY, Thor, Olivia, etc.

Using this program, it is possible to communicate worldwide using only a few watts of RF power. The program can also be used for amateur radio emergency communications when other communication systems fail, e.g., due to natural disasters or power outage.

Fldigi can run on many platforms, such as Windows, macOS, Linux, FreeBSD, OpenBSD, Solaris etc. Multiple sound systems are supported by fldigi, such as Open Sound System (OSS), PortAudio, PulseAudio, etc.

13.2 Features

Some important features of fldigi are:

- Support of transmission and reception in all languages (using UTF-8 character)
- Narrowband emergency messaging system
- Connection to outside via TCP/IP on port 7322
- DTMF encoding and decoding
- Can be used as a KISS modem via TCP port 7342
- Sound card oscillator frequency correction
- Measure of RF receiver frequency skew and sound card oscillator skew
- Control of external radio hardware
- Simultaneous decoding of multiple Morse code (CW) signals

13.3 Digital Formats

Fldigi supports many digital formats. Some popular ones are:

Morse code	5 – 50 words per minute
PSK	31, 63, 63F, 125, 250, 500, 1000
FSQ	2, 3, 4, 5, 6
DominoEX	Micro, 4, 5, 8, 11, 16, 22, 44, 88
MFSK	4, 8, 11, 16, 22, 31, 32, 64, 64L, 128, 128L
QPSK	31, 63, 125, 250, 500
NAVTEX	Navtex/SitorB
RTTY	45, 45/170, 50/170, 75.170, 75.850
WEFAX	IOC-576, IOC-288

The following YouTube video gives a good introduction to FLDIGI:

https://www.youtube.com/watch?v=jvOJFFkYlAs

Figure 13.1 shows the typical FLDIGI based system setup. Audio output from the transceiver is sent to the Raspberry Pi 5 audio adapter microphone input, and the speaker output port of the audio adapter is connected to the line input of the transceiver.

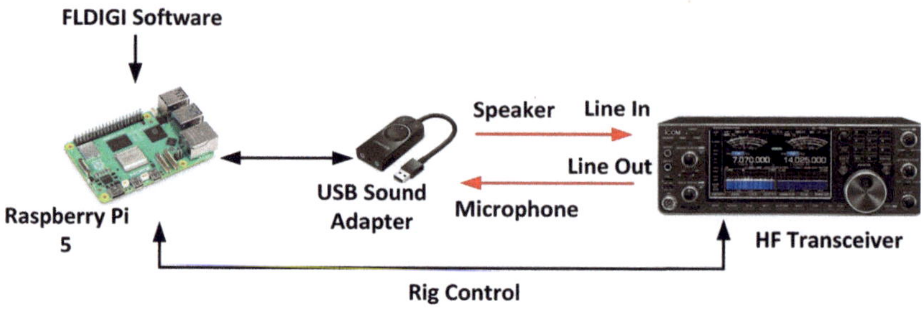

Figure 13.1: FLDIGI based system setup.

13.4 Installation on Raspberry Pi 5

This section discusses the installation of **fldigi** from the "Raspbian" repository. The steps are given below:

- Start the Desktop GUI.
- Click the **Applications Menu** and then click to open **Preferences** and select **Add/Remove Software.**
- Type **fldigi** and press Enter as shown in Figure 13.2.
- Select digital modem program for ham radio operators and click Apply to install the software.
- Click OK and exit from the menu.

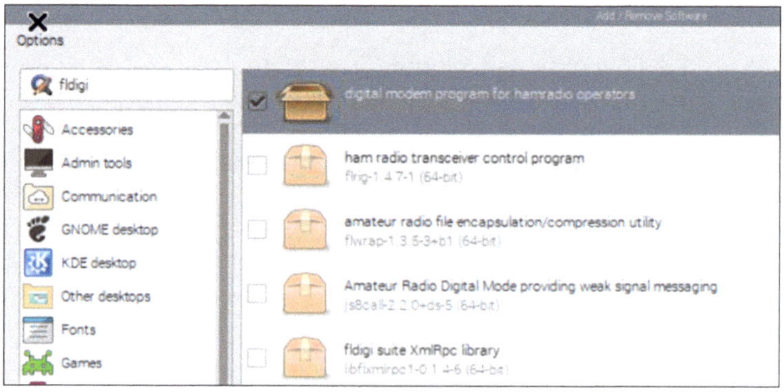

Figure 13.2: Select the program and click Apply.

13.5 Starting the Program

To start the program, make sure you are in Desktop mode, click **Applications Menu → Hamradio → fldigi**. When the program is run for the first time, you will be presented with a configuration wizard as shown in Figure 13.3.

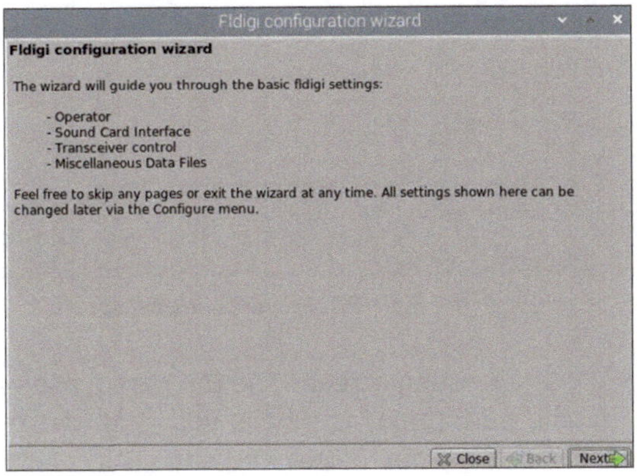

Figure 13.3: The Fldigi screen.

Click **Next** and enter the details such as the station name, operator callsign, station QTH etc. (Figure 13.4). Click **Next** to configure the audio. Click **Devices** and then **PulseAudio** and select **Default** and click **Enable** next to it as shown in Figure 13.5. Click **Finish** to exit the configuration menu. You may also want to configure your rig either during the first startup or by clicking the **Configure** menu and selecting **Rig Control → Rig → Hamlin** after the program is up and running.

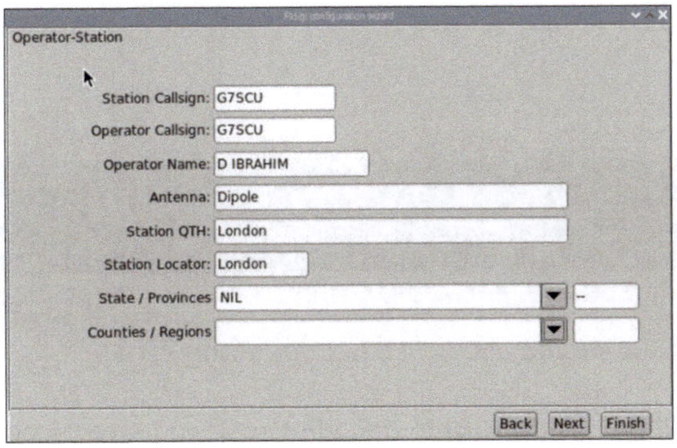

Figure 13.4: Enter callsign, station, etc.

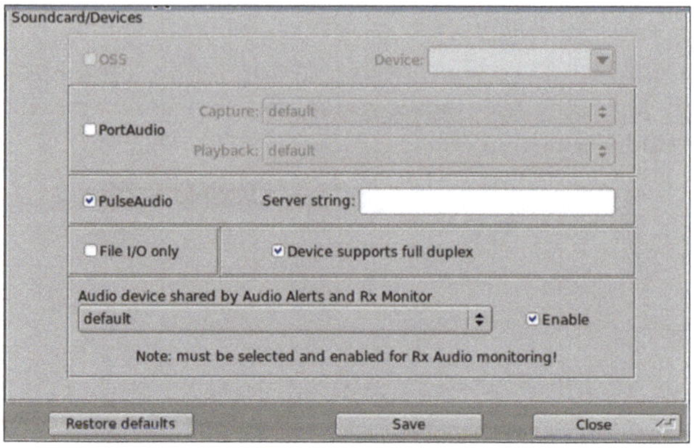

Figure 13.5: Enter audio details,

Figure 13.6 shows the FLDIGI operating screen. The top part (yellow) is where the received text is displayed. The middle part (blue) is where you type the text to be transmitted. Under the middle part are some buttons where you can click to activate various controls. The operation mode is selected by clicking menu item **Op Mode** at the top left part of the screen.

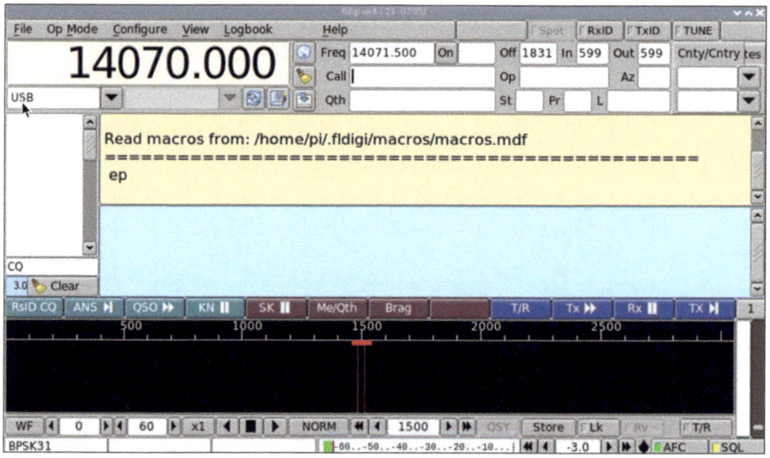

Figure 13.6: FLDIGI operating screen.

In this chapter you will be using **fldigi** to decode Morse code, to receive and display WEFAX weather messages, to receive and display RTTY and NAVTEX messages.

13.6 Decoding Morse Code (CW)

In this section you will receive Morse code in audio format and then decode and display it on the fldigi screen. For simplicity, let's use a PC to generate test Morse code, but you can use your receiver hardware tuned to a CW transmission for this purpose.

The steps are as follows (see Figure 13.7):

- Connect the audio output of your PC to the microphone input of the Raspberry Pi 5 through a USB audio adapter (e.g. UGREEN).
- Start the fldigi program.
- Select **CW** at the top left and then select **CW** in the **OP Mode**.
- There are many sample Morse code audio files on the Internet (e.g., see website https://commons.wikimedia.org/wiki/Category:Audio_files_of_Morse_code). Play one of the files (e.g. **A through Z in Morse code**), making sure that the audio volume is high enough, and click in the waterfall to start the decoding. You should see the signal level meter (in green) at the center bottom part of the screen moving to show the received signal levels.
- You should see the decoded Morse code text displayed on the yellow part of your fldigi screen as shown in Figure 13.8.

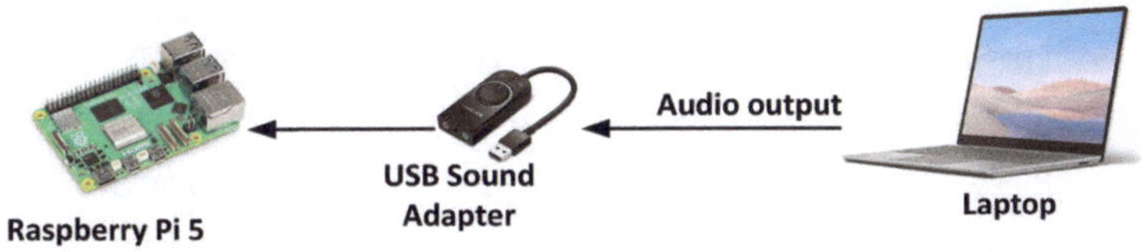

Figure 13.7: Stup to receive test Morse code.

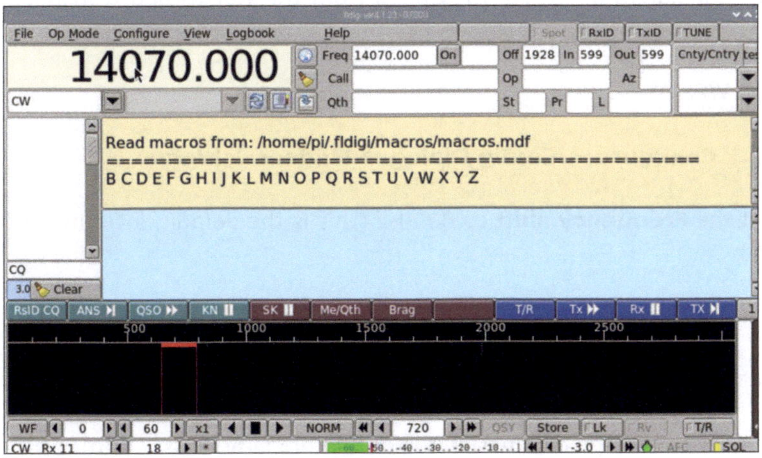

Figure 13.8: Decoded Morse code letters.

13.7 Receiving Weather Fax (WEFAX)

WEFAX (or Weatherfax; Weather Facsimile) is a slow-scan image transmission of weather charts and meteorological reports.

WEFAX is transmitted at 60, 90, 100, 120, 180 or 240 LPM (Lines per Minute) speeds with modes called IOC 576 or IOC 288. Most weather forecasts are sent at 120 LPM in IOC 576. In this format, the duration of the fax is as follows:

Start tone:	5 seconds
Phasing signal:	30 seconds
Image:	10 minutes (at 120 LPM)
Stop tone:	5 seconds
Black:	10 seconds

At 120 LPM, fax is transmitted line by line at a rate of 120 lines per minute (or half a second per line). The pixels in the image are converted into certain audio tones. For example, 1500 Hz represents black, 2300 Hz represents white, and frequencies in between represent various shades of gray.

Weather faxes are transmitted by many countries around the world. For example, Northwood in UK (GYA Northwood) transmits weather fax messages at specified times (see link: http://www.yachtcom.co.uk/comms/weather/) on the following frequencies:

Nominal frequency	GQRX freq	Times
2618.5 kHz	2616.6 kHz	2000-0600 UTC
4610.0 kHz	4608.1 kHz	0000-2400 UTC
8040.0 kHz	8038.1 kHz	0000-2400 UTC
11086.5 kHz	11084.6 kHz	0600-2400 UTC

You can set your fldigi for WEFAX messages from the Northwood transmitter format as follows:

Configure the WEFAX:

- Click **Configure → Config Dialog → Modem → Wefax**
- Set the **Frequency shift** to 800 Hz (this is the default, inGermany this is 850 Hz)
- Set the **Center frequency** to 1900 Hz
- Set the **Filter** to Medium
- Specify the image destination folder and filename. e.g. **/home/pi/**
- Leave the other settings as shown in Figure 13.9. Click **Save** and then **Close**

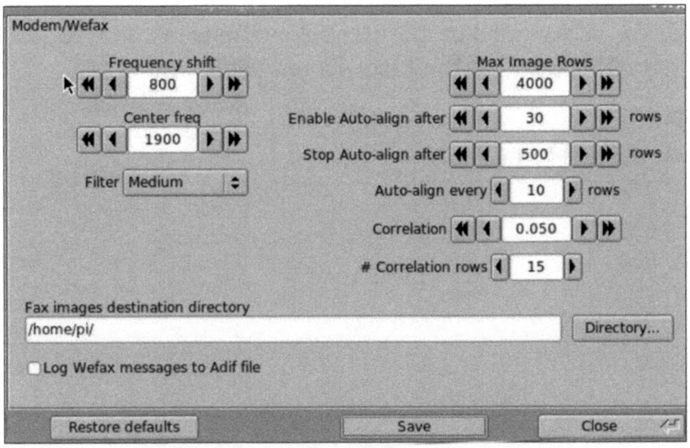

Figure 13.9: WEFAX configuration.

The definitions of the WEFAX configuration settings are as follows:

Frequency shift adjustment: the default is 800 Hz (850 Hz for the Deutsche Wetter Dienst).
Center frequency: the center frequency of the USB modulated signal.
Maximum rows: the image is automatically saved when it has more than this number of rows.
Auto align: the decoder searches for the start, stop, and phasing components of the transmission.

The buttons at the bottom of the screen are:

Save: saves the current image as a PNG file
Clear: clears the image
Cont': the image can be paused and then resumed
Mag: changes the displayed magnification of the Wefax image
Tilt: used to adjust the slant of the image A value of 0.0071 should be satisfactory (set to 0.0118 for Germany)
Align: used to manually adjust the horizontal center of the image
Auto: press to center the image
Noise: used to enable noise filtering.
Bin: convert to binary image using the specified level

Receiving WEFAX:

- Start the fldigi program.
- Click **OP Mode** and select **WEFAX** and then **WEFAX-IOCIOC756**.
- Set the frequency to 4608.1 kHz with the nominal frequency at 4610.0 kHz.

- Tune your wireless radio to 1.9 kHz below the transmit frequency, i.e. at 4608.1 kHz. Figure 13.10 shows the fldigi screen setup

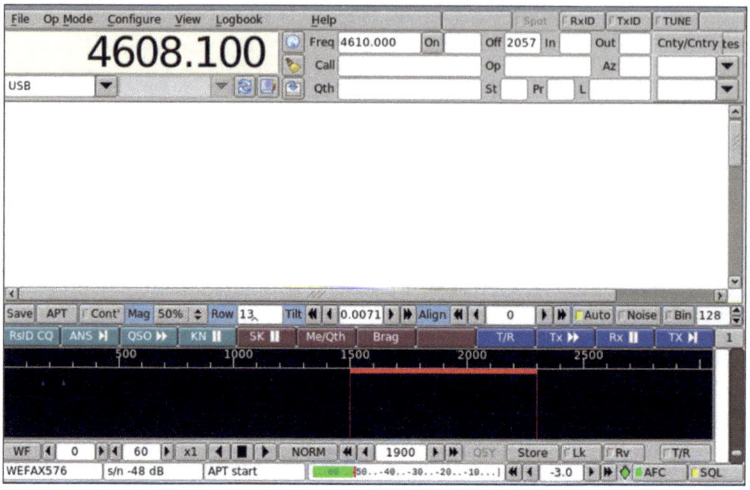

Figure 13.10: fldigi screen setup.

- Check the time of transmission and wait to receive the WEFAX data.

- at the time of writing the book, a sample fax sound file was available at the following link:

  https://www.youtube.com/watch?v=FRfCYKQQ3_4

- You should align the two vertical red lines at the edges of the waterfall.

- Notice that the image is built in real time on the fldigi screen as data is received. The row number is increased and the signal meter moves to indicate data reception. The row counter will stop when the data reception is complete. Figure 13.11 shows the fldigi screen as the data is received.

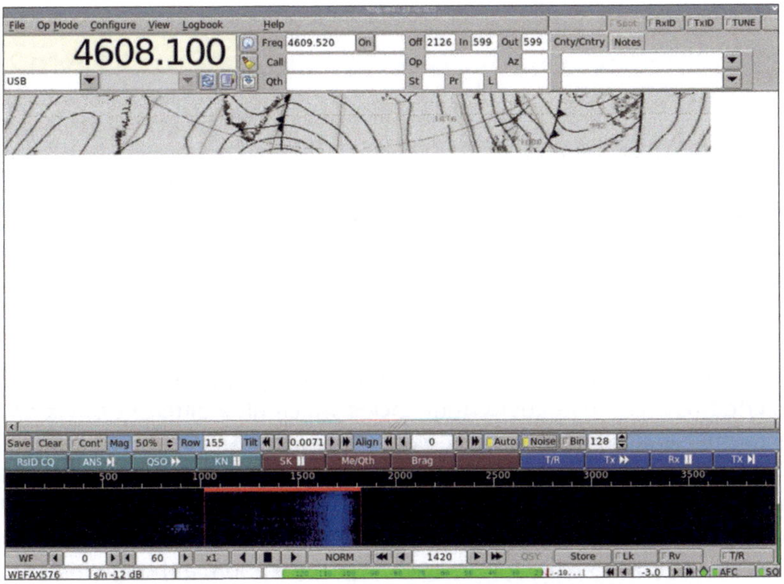

Figure 13.11: The screen as the data is received.

The image is stored in the specified directory in the following format:

wefax_yyyymmdd+hhmmss_fffffff_nocorr.png

For example, if the image was received on the 11th of February, 2024 at 21:50:45 with the frequency set to 4608.100 Hz, then the filename will be:

wefax_20240211_21:50:45_4608100_nocorr.png

The received image is shown in Figure 13.12.

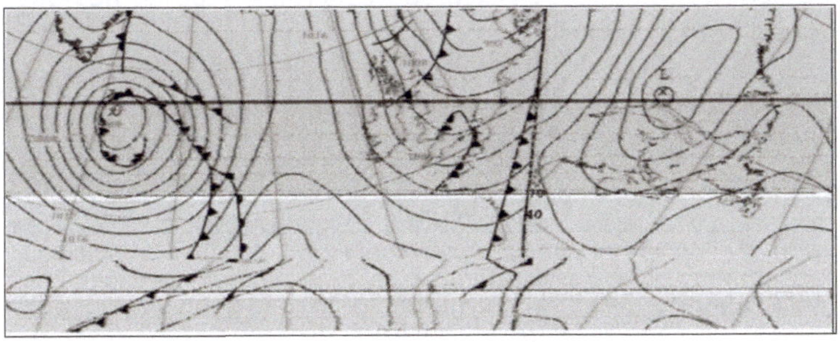

Figure 13.12: Received image.

The following weather fax related websites can be of interest to readers:

http://www.bbrc.info/articles/receiving-weatherfax-maps
http://www.blackcatsystems.com/radio/rfax.html

http://www.blackcatsystems.com/software/multimode/fax.html#GFA
http://www.blackcatsystems.com/software/multimode/fax.html#CFH
http://www.blackcatsystems.com/software/multimode/fax.html#NMC

13.8 Receiving RTTY Traffic

RTTY (Radio TeleTYpe) is a method of using tones to send digital messages between stations oparting in amateur HF bands. Some other services also still employ RTTY. You can send messages with RTTY using several coding methods such as Baudot, Amtor, simple ASCII, etc. You will need to modulate the characters using AFSK (audio frequency shift keying) or FSK (frequency shift keying).

The code representing letters consists of low and high voltages (known as Mark and Space) which are shifted between two audio frequencies which give different tones on a speaker. The frequency shift is around 200 Hz. A Mark represents 1445 Hz and a Space 1275 Hz. Changing to upper case (or vice versa) is done using shift keys. RTTY has the disadvantage that it can be used to send only letters, characters and a few other symbols. Data is sent relatively slowly because the old "teleprinters" couldn't achieve fast operation. The standard RTTY speed is 45.5 baud for amateur radio operation, although faster speeds such as 50, 75, 88 are also available.

RTTY operation is usually found in the following bands:

160 meters, between 1.800 and 1.820 MHz
80 meters, between 3.58 and 3.65 MHz
30 meters, between 10.110 and 10.150 MHz
20 meters, between 14.080 and 14.099 MHz
15 meters, between 21.080 and 21.100 MHz
10 meters, between 28.080 and 28.100 MHz

RTTY is used by ships to receive weather information. Across Western Europe and the Middle East, the German Weather Service broadcasts RTTY weather messages for mariners at the frequencies shown in Table 13.1.

Table 13.1: RTTY weather messages at Western Europe and Middle East (https://weather.mailasail.com/Franks-Weather/Radio-Teletype-Weather-Broadcasts)

Frequency	Callsign	Times of broadcast	Power	Class of emission	Speed	Shift
147.3 kHz	DDH 47	05.30 - 22.00 UTC	20 kW	F1B	50 Baud	± 42.5 Hz
11039 kHz	DDH 9	05.30 - 22.00 UTC	1 kW	F1B	50 Baud	± 225 Hz
14467.3 kHz	DDH 8	05.30 - 22.00 UTC	1 kW	F1B	50 Baud	± 225 Hz
4583 kHz	DDK 2	00.00 - 24.00 UTC	1 kW	F1B	50 Baud	± 225 Hz
7646 kHz	DDH 7	00.00 - 24.00 UTC	1 kW	F1B	50 Baud	± 225 Hz
10100.8 kHz	DDK 9	00.00 - 24.00 UTC	10 kW	F1B	50 Baud	± 225 Hz

13.8.1 Using fldigi to receive RTTY messages

An example is shown is this section how RTTY messages can be received and displayed using the fldigi program. At the time of writing this book, a demo site was available on the Internet at the following link that generates RTTY messages for test purposes:

https://www.youtube.com/watch?v=wzkAeopX7P0

The steps are as follows:

- Start the fldigi program.
- Click **OP Mode** and select **RTTY** and set the speed to 45.5 baud (RTTY-45).
- If you are using the above test RTTY code, go to the specified web site. Connect the audio output of your laptop to microphone input of your Raspberry Pi 5 audio adapter.
- Start to play the RTTY code on your laptop.
- Set the two vertical lines to the sides of the waterfall. You should see messages similar to Figure 13.13 displayed on your fldigi screen.
- As an exercise, you may want to try to tune to a real RTTY transmission.

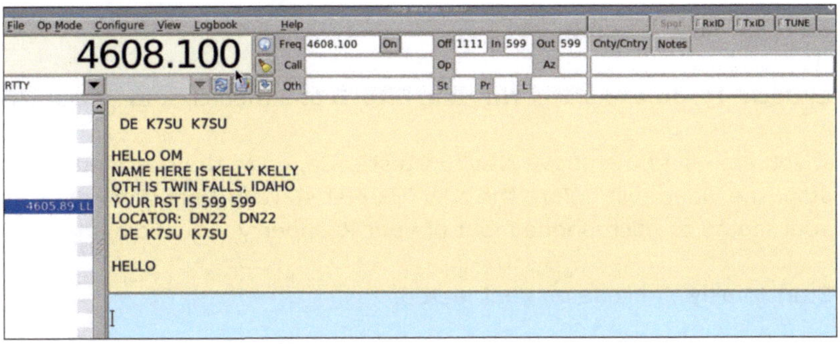

Figure 13.13: Test RTTY.

13.9 Receiving NAVTEX Messages

NAVTEX, or **NAV**igational **TE**le**X**, is an international medium frequency (518 kHz and 490 kHz) service for broadcasting free navigational and meteorological warnings and forecasts as well as safety information to ships. Safety information to ships (MSI) is also broadcast on 4209.5 kHz.

NAVTEX messages are transmitted in BFSK (Binary Frequency Shift Keying) at 100 bits per second and with 170 Hz phase shifting. A list of European NAVTEX stations can be found on the following web site:

https://weather.mailasail.com/Franks-Weather/European-And-Mediterranean-Navtex-Schedules

Warnings to ships are broadcast as soon as possible and repeated every 4 hours. Warnings are listed by areas with a gale warning in force.

There are 3 transmitting stations in the UK and 2 stations in Ireland. A total coverage of the UK and Irish waters are covered with these stations. The stations are given unique identification letters so that ships can identify them.

13.9.1 Using fldigi to receive NAVTEX messages

An example is shown is this section how NAVTEX messages can be received and displayed using the fldigi program. At the time of writing this book, a test site was available on the Internet at the following link that generates weekly NAVTEX messages in a YouTube video for testing the FURUNO NAVTEX hardware. The message generates text "THIS IS AN INTERNAL TEST MESSAGE" and then generates code for letters of the alphabet, numbers and some symbols. The actual NAVTEX messages start from about mid-point of the video. Listen to the video and find the point where the NAVTEX messages start. Stop the video just before this point:

https://www.youtube.com/watch?v=5KntVVyvewI

The steps are as follows:

- Start the fldigi program.
- Click **OP Mode** and select **Navtex/SitorB** and then click to select NAVTEX.
- If you are using the above NAVTEX test code, go to the specified web site and stop. the video just before the NAVTEX test starts. Connect the audio output of your laptop to microphone input of your Raspberry Pi 5 audio adapter.
- Start to play the code on your laptop.
- Set the two vertical lines to the sides of the waterfall. You should see messages similar to Figure 13.14 displayed on your fldigi screen.
- As an exercise, you may want to try to tune to a real NAVTEX transmission.

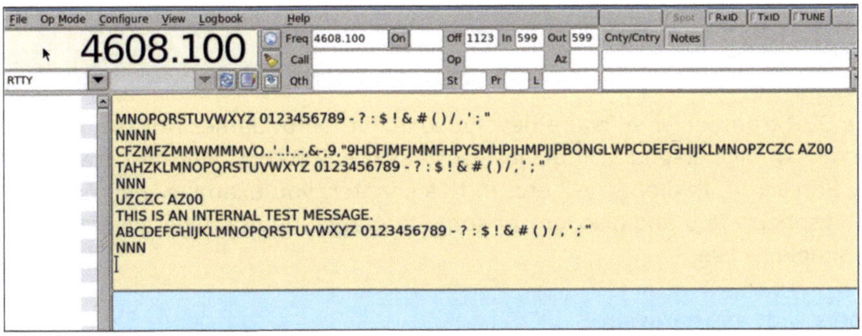

Figure 13.14: Give NAVTEX decoding a test run.

Chapter 14 • Quisk

14.1 Overview

Quisk is a SDR transceiver software developed in Python by James Ahlstrom (N2ADR). The software runs on higher-end SDR devices such as the SoftRock, Hermes Lite-2, Red Pitaya, Odyssey, SdrMicron, RadioBerry-2 etc. In this chapter, you examine the installation of the quisk on Raspberry Pi 5 and also look at some features of quisk. One advantage of quisk is that it is simple to use.

Quisk works with this hardware:

- SoftRock connected to the sound card.
- many other SDRs connected to the sound card.
- SDR-IQ connected by USB.
- Perseus connected by USB.
- N2ADR hardware connected by Ethernet and IP.
- HiQSDR hardware connected by Ethernet and IP.
- the Hermes-Lite project at hermeslite.com.

Quisk is written in Python and C and the source code is available. Data can come from a sound card, Ethernet or USB. Quisk receiver can read data, filter it, demodulate it, and send the audio to the sound card for output to speakers or headphones. The quisk transmitter accepts a microphone input and sends to a transmitter via a soundcard or Ethernet.

14.2 Installing quisk on the Raspberry Pi 5

Quick can easily be installed on the Raspberry Pi 5 running in Desktop mode:

- Start the Desktop

- Click **Applications Menu → Preferences → Add/Remove Software**.

- Enter **quisk** in the Search box and press Enter.

- Select **Software Defined radio (SDR)** and click **Apply** to install.

- Enter your Raspberry Pi 5 password.

- Wait until the software is installed.

To run quisk, open a terminal session in Desktop and enter the command quisk. Alternatively click **Applications Menu → Hamradio → quisk**:

pi@raspberrypi:~ $ **quisk**

Figure 14.1 shows the quisk startup screen. Click the **Config** button and then click **Radios** to select your radio type as shown in Figure 24.28.

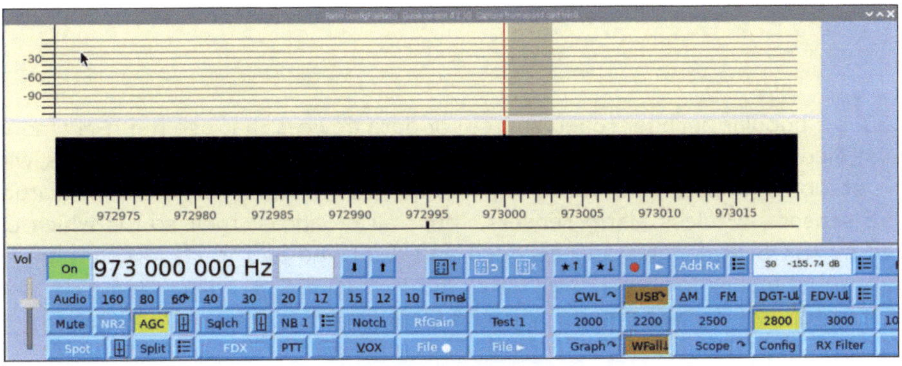

Figure 14.1: Quisk startup screen.

Figure 14.2: Select your radio.

Give a name to your radio (at the right hand side) and click **Add** (you may have to move the slider to the right). Click **Config** followed by **Help with Radios** to display help on using quisk.

More information on quisk can be obtained from the following web pages:

https://james.ahlstrom.name/quisk/
https://community.linuxmint.com/software/view/quisk
https://www.oz9aec.net/radios/gnu-radio/quisk-a-software-defined-radio-for-linux
https://www.youtube.com/watch?v=yADRRJAJscg

Chapter 15 • RTL_433

15.1 Overview

RTL_433 is a popular decoder (open source) of data devices that use the ISM (Industrial, Scientific, Medical) bands to broadcast data wirelessly. Low-cost consumer devices, wireless weather stations, wireless devices such as wireless thermostats, wireless temperature and pressure sensors, wireless alarm sensors, etc., all broadcast their status which can be received by RTL-SDR dongles and decoded and displayed by the **rtl_433** program. The program works mainly for the 433.92 MHz, 868 MHz (SRD), 315 MHz, 345 MHz, and 915 MHz ISM bands. The default frequency is 433.92 MHz with 250 kHz sample rate.

15.2 Installing RTL_433 on the Raspberry Pi 5

Enter the following command to install tl_433 on Raspberry Pi 5:

pi@raspberrypi:~ $ **sudo apt-get rtl-433**

Run the program by simply typing **rtl_433**. The program initially displays the messages shown in Figure 15.1 when it is started.

```
pi@raspberrypi:~ $ rtl_433
rtl_433 version 22.11 (2022-11-19) inputs file rtl_tcp RTL-SDR SoapySDR
Use -h for usage help and see https://triq.org/ for documentation.
Trying conf file at "rtl_433.conf"...
Trying conf file at "/home/pi/.config/rtl_433/rtl_433.conf"...
Trying conf file at "/usr/local/etc/rtl_433/rtl_433.conf"...
Trying conf file at "/etc/rtl_433/rtl_433.conf"...
Registered 191 out of 223 device decoding protocols [ 1-4 8 11-12 15-17 19-23
-26 29-36 38-60 63 67-71 73-100 102-105 108-116 119 121 124-128 130-149 151-16
163-168 170-175 177-197 199 201-215 217-223 ]
Found Rafael Micro R828D tuner
RTL-SDR Blog V4 Detected
Exact sample rate is: 250000.000414 Hz
Sample rate set to 250000 S/s.
Tuner gain set to Auto.
Tuned to 433.920MHz.
baseband_demod_FM: low pass filter for 250000 Hz at cutoff 25000 Hz, 40.0 us
```

Figure 15.1: Initial messages.

A sample default display is shown in Figure 15.2 with no options specified. This shows the temperature, humidity battery status, pressure etc. of various devices transmitting close to the RTL-SDR dongle antenna. It is clear that some of the data comes from nearby vehicles and temperature and humidity sensors.

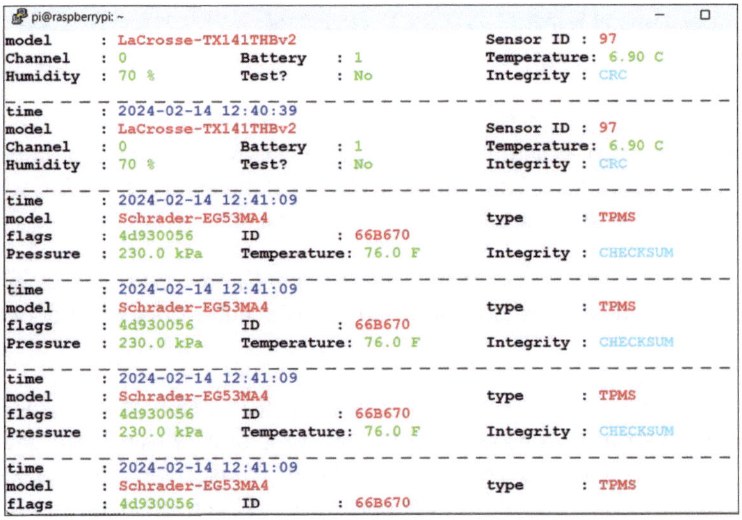

Figure 15.2: Default display.

15.3 Program Options

The program options can be displayed by entering the command:

pi@raspberrypi:~ $ **rtl_433 -h**

Some commonly used options are:

-f :	receive frequency. The default is: 433920000 Hz
-g <gain>:	the default is auto
-V:	display the version
-r <filename>:	read data from file instead of a receiver
-w <filename>:	save data to a file
-W <filename>:	save data to a file, overwrite existing data
-n <value>:	specify number of samples to take
-T <seconds>:	specify number of seconds to run

You can configure to start the RTL_433 at boot time if you wish.

Chapter 16 • Other SDR hardware

16.1 Overview

So far, we have been using the RTL-SDR type dongles with SDR programs. For most general purpose applications, such as listening to the broadcast stations, amateur bands, air traffic, monitoring satellites, and so on, the cheap RTL-SDR dongles are great because the signal levels are usually high and indoor aerials can be used.

There are many other SDR hardware in the form of USB dongles that are available in the market and can be used with the SDR programs. This chapter presents a brief look at some of these devices.

Some SDRs like HackRF One, Red Pitaya, RS-HFIQ, Xiegu G90, RS-918, Hermes Lite-2, and others are both RX and TX devices. One of the problems of using SDR devices with the Raspberry Pi 5 is the requirement of high CPU speed. Transceiver software like Quisk, and linHPSDR is available for TX/RX type SDR hardware and can be installed on a Raspberry Pi 5.

16.2 HackRF One

This SDR transceiver (web link: https://greatscottgadgets.com/hackrf/one/) from Great Scott Gadgets costs around £320, and it has the following features (Figure 16.1).

- 1 MHz to 6 GHz operating frequency
- 20 million samples/s
- Half-duplex TX/RX
- Compatible with popular software (SDR#, GNU radio etc)
- USB 2.0 interface
- USB powered
- 8-bit resolution
- Open source hardware
- SMA female antenna connector
- Protected by molded plastic enclosure

Figure 16.1: The HackRF One.

The transmit power is up to 10 dBm for the frequency range up to 4 GHz. An external RF amplifier is recommended to increase the power if required. The maximum receive power is -5 dBm.

Opera Cake (Figure 16.2) is an antenna switching add-on board for HackRF One that is configured with command-line software either manually, or for automated port switching based on frequency or time. It has two primary ports, each connected to any of eight secondary ports, and is optimized for use as a pair of 1×4 switches or as a single 1×8 switch. Its recommended frequency range is 1 MHz to 4 GHz.

When HackRF One is used to transmit, Opera Cake can automatically route its output to the appropriate transmit antennas, as well as any external filters and amplifiers. No changes are needed to the existing SDR software, but full control from the host is available.

Opera Cake also enhances the HackRF One's use as a spectrum analyzer across its entire operating frequency range of 1 MHz to 4 GHz. Antenna switching works with the existing hackrf_sweep feature, which can sweep the whole tuning range in less than a second. Automatic switching mid-sweep enables the use of multiple antennas when sweeping a wide frequency range.

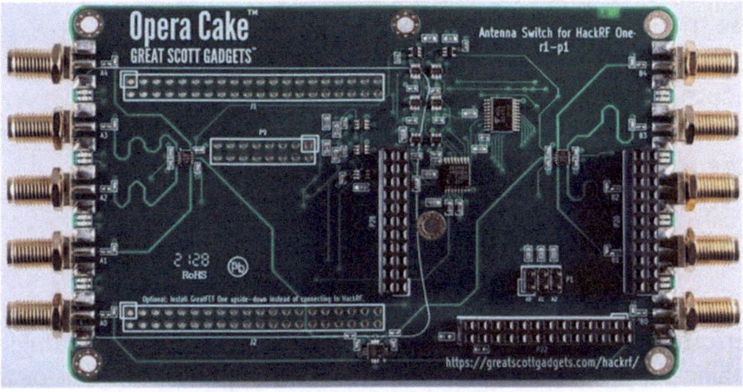

Figure 16.2: Opera Cake.

HackRF One is sold as a bundle including the following items:

- 1× HackRF One SDR
- 1x Injection molded plastic enclosure
- 1x micro-USB cable
- 1x Opera Cake Antenna Switch

16.3 NooElec NESDR Smart HF Bundle

The NooElec NESDR Smart HF is a 100 kHz – 1.7 GHz SDR bundle (Figure 16.3) for HF/UHF/VHF, including RTL-SDR, upconverter, balun, all necessary adapters, desktop indoor antennas, and cables, costing £103. The bundle is a good starting kit for those who wish to enter into the world of SDR. Using the upconverter, the RTL-SDR can operate at as low as 100 kHz, thus enabling the amateurs to use the HF.

Figure 16.3: The NooElect SDR.

The instructions to use the SDR in the HF band are as follows:

- Connect the antenna to the Ham It Up unit.
- Connect the SDR to the IF output using the supplied connector.
- Make sure the toggle switch is in the Enable position.
- Connect the SDR to your Raspberry Pi using a USB cable.
- Plug the Ham It Up USB-B USB jack to a USB power source.
- Start using the device. Tune to 125 MHz (+/– the tuning offset from the tuning procedure) + your desired frequency.

16.4 BladeRF

This is a high performance SDR receiver/transmitter (Figure 16.4) and unlike the HackRF, it is full duplex, making it ideal for higher performance applications. Out of the box BladeRF can tune from 300 MHz to 3.8 GHz.

Figure 16.4: BladeRF.

Some of its salient features are:

- Independent RX/TX 12-bit 40 MSPS quadrature sampling
- 16-bit DAC factory calibrated
- On-board 200 MHz ARM9 processor with 512 KB SRAM
- Fully bus powered
- DC power jack for running headless
- Extensible gold plated RF SMA connectors
- Linux, Windows, Mac and GNURadio software support
- The hardware can be operated as a spectrum analyzer, vector signal analyzer, and vector signal generator

The Blade RF 2.0 micro xA5 has the frequency range 47 MHz to 6 GHz, costing around $670.

16.5 LimeSDR

This device (Figure 16.5) is capable of receiving and transmitting UMTS, LTE, GSM, LoRa, Bluetooth, ZigBee, RFID, digital broadcasting and more. The device is integrated into the Snappy Ubuntu core and can be used through an app.

Figure 16.5: LimeSDR.

Some features of the LimeSDR are:

- RF transceiver
- 256 MB memory
- FPGA Altera Cyclone
- USB 3.0 controller
- 100 kHz to 3.8 GHz continuous frequency range
- 61.44 MHz bandwidth
- 10 dBm power output (CW)
- microUSB or external power supply
- There is also a smaller less expensive version of the LimeSDR known as the LimeSDR Mini.

16.6 Universal Software Radio Peripheral (USRP)

USR (Figure 16.6) was developed by Ettus Research and has been very popular especially among academic researchers, allowing connection of the device to a host computer and control hardware for data transmission/reception.

Some features of the USRP are:

- frequency range 70 MHz to 6 GHz
- 12-bits ADC and DAC
- 61.44 MHz bandwidth
- Xilix Spartan-6 FPGA

Figure 16.6: USRP.

16.7 ADALM-Pluto

ADALM-Pluto (Figure 16.7) is an easy to use SDR, made as a learning aid for teachers and students, to impart the fundamentals of the software defined radio. Its basic features are:

- Frequency range: 325 MHz to 3.8 GHz
- 12-bit ADC and DAC
- 20 MHz channel bandwidth
- Xilinx Zynq Z-7010 FPGA
- MATLAB and Simulink support
- USB 2.0 powered

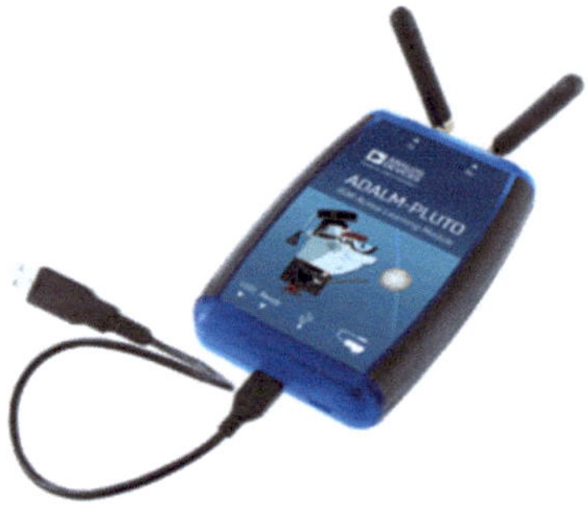

Figure 16.7: ADALM-Pluto.

16.8 AirSpy HF+ Discovery

This is a high performance SDR (Figure 16.8) developed jointly between Airspy, Itead Studio, and ST Microelectronics (see website: https://airspy.com/airspy-hf-plus/).

The basic technical specifications of this device are:

- HF coverage starting from 9 kHz
- 60 to 260 MHz VHF coverage (new receivers also cover 0.5 kHz to 31 MHz)
- 22-bit resolution
- 0.5 ppm high precision low noise clock
- 1 ppb frequency adjustment
- Excellent noise reduction (claimed to be the best one in the market)
- High dynamic range (ADC up to 36 MSPS)
- Wide band RF filter bank
- Tracking RF filters
- Inputs matched to 50 ohms
- 4× GPIO
- Smart AGC with real time optimization
- Supported by well-known software, such as SDR Sharp, SDR-Console, GQRX, HDSR, Krypto1000, etc.
- Supported by operating systems, such as Windows, Linux, BSD, OSX
- Supported by hardware, such as Windows PPC, Raspberry Pi, Odroid, etc

Figure 16.8: AirSpy HF+ Discovery.

In 2019, the Airspy HF+ Dual Port won the prestigious WRTH Award in front of a fierce competition of highly refined communication receivers and other SDRs. The device offers an excellent noise reduction algorithm.

Chapter 17 • Installation and Use of Some Popular Radio Applications

17.1 Overview

In this chapter you will learn how to install and use some other amateur radio programs.

17.2 Aldo Morse Code Tutor – Text Based

Aldo is a text based Morse Code training program. The program offers several training methods. Morse code is output on the speaker by default.

17.2.1 Installing aldo on the Raspberry Pi 5

The steps are:

- Start the Desktop.
- Click **Applications Menu → Preferences → Add/Remove Software**.
- Enter **aldo** in the search box and press Enter.
- Enter your Raspberry Pi 5 password.
- Wait until the software is installed.

To run aldo, open a terminal session in Desktop and enter the command **aldo**. The program will start with the menu options shown in Figure 17.1. Choose an option and you should hear the Morse code output on your speaker:

```
pi@raspberrypi:~ $ aldo
Aldo 0.7.7 Main Menu
        1: Blocks method
        2: Koch method
        3: Read from file
        4: Callsigns
        5: Setup
        6: Exit
Your choice:
```

Figure 17.1: aldo menu options.

17.3 xcwcp Morse Code Tutor — Graphical

Aldo is a graphical Morse Code training program. The program offers several training methods. Morse code is output on the speaker by default.

17.3.1 Installing xcwcp on Raspberry Pi 5

The steps are:

- Start the Desktop.
- Click **Applications Menu → Preferences → Add/Remove Software**.

- Enter **xcwcp** in the search box and press Enter.
- Enter your Raspberry Pi 5 password.
- Wait until the software is installed.

To run xcwcp, start your Desktop GUI, click **Application Menu → Hamradio** and then click on **xcwcp**. Alternatively, open a terminal session n Desktop mode and enter the command xcwcp. As shown in Figure 17.2, you will be presented with a screen. Select the type of output you want (e.g., English words), speed (wpm), tone (e.g. 800 Hz), volume (say, 70%) and the gap. Click the green arrow button at the top left of the screen to start generating Morse code. You should hear the output on your speaker.

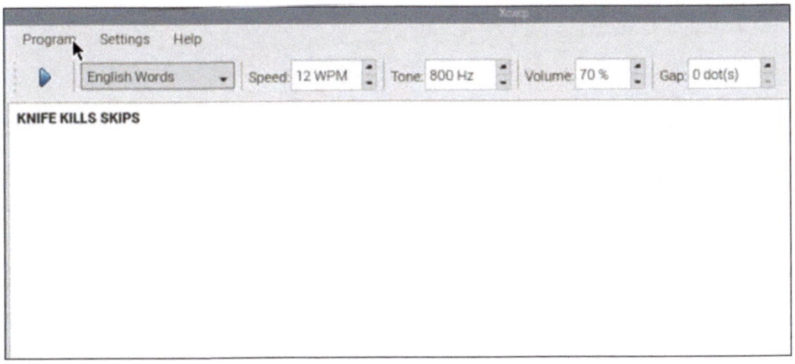

Figure 17.2: xcwcp screen.

17.4 GPredict — Satellite/Orbital Object Tracking and Rig Control

GPredict is a real-time program tracking satellites, predicting their positions and velocity at a given time, and also can do rig control. There is no software limit on the number of satellites being tracked. Satellite data is given in the forms of maps and tables and polar plots. Satellites can be grouped into modules for simplicity. Support os provided for automated ground station operations providing both Doppler tuning for radios and anrenna rotator control. Further details on the GPredict program are available in the following document:

> *GPredict User Manual, Work in progress for GPredict 2.x,*
> by Alexandru Csete, OZ9AEC.

17.4.1 Installing GPredict on Raspberry Pi 5

To install GPredict, follow the steps given in in the previous sections, enter GP redict to search box, and wait until the program is installed.

Start the program in Desktop mode by selecting **Applications Menu → Hamradio** and click on **Gpredict**. Alternatively open the terminal in Desktop mode and enter the command: **GPredict**. Figure 17.3 shows the startup menu.

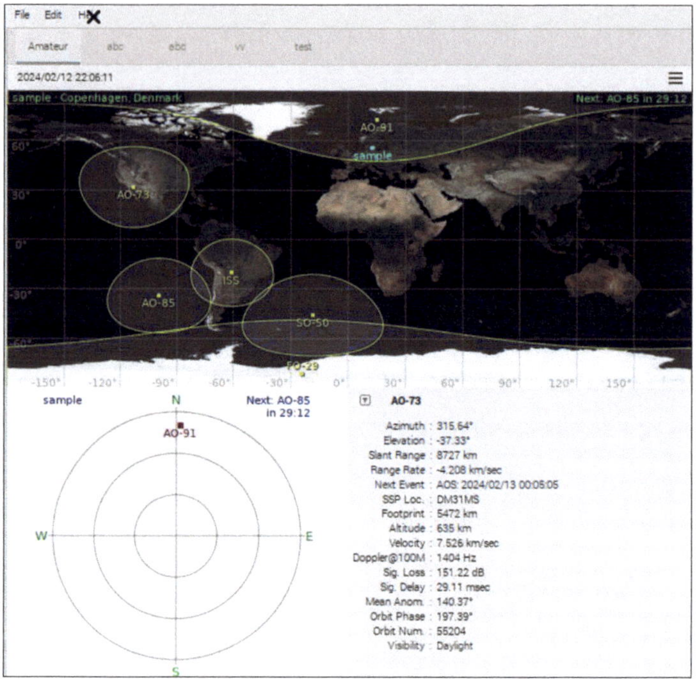

Figure 17.3: GPredict startup menu.

17.4.2 GPredict controls

Selecting a satellite:

- Click on **File → New Module**
- Choose a satellite. e.g., **GPS Operational**, GPS BIIF-1 (PRN 25). Double-click on the satellite name at the left pane or click on the satellite name and click the right arrow to move the satellite name to the right pane (Figure 17.4).
- Choose a **Module Name**, e.g. **FirstGPS**.
- Click **OK**.
- Figure 17.5 shows the map and the selected satellite. Navigational data is given about the selected satellite at the bottom right hand side of the map.

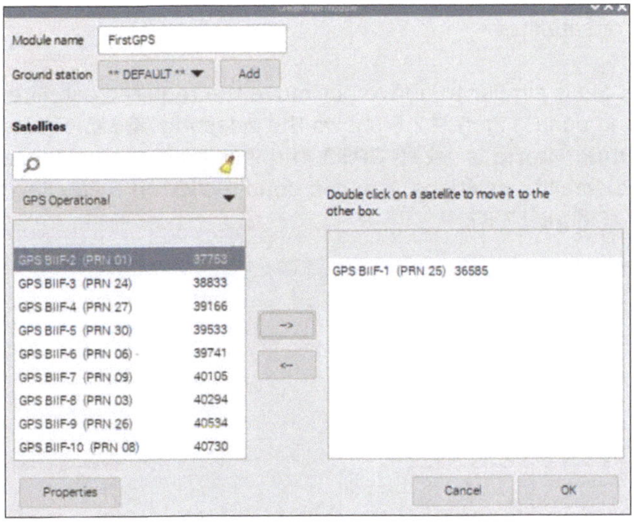

Figure 17.4: Selecting a GPS satellite.

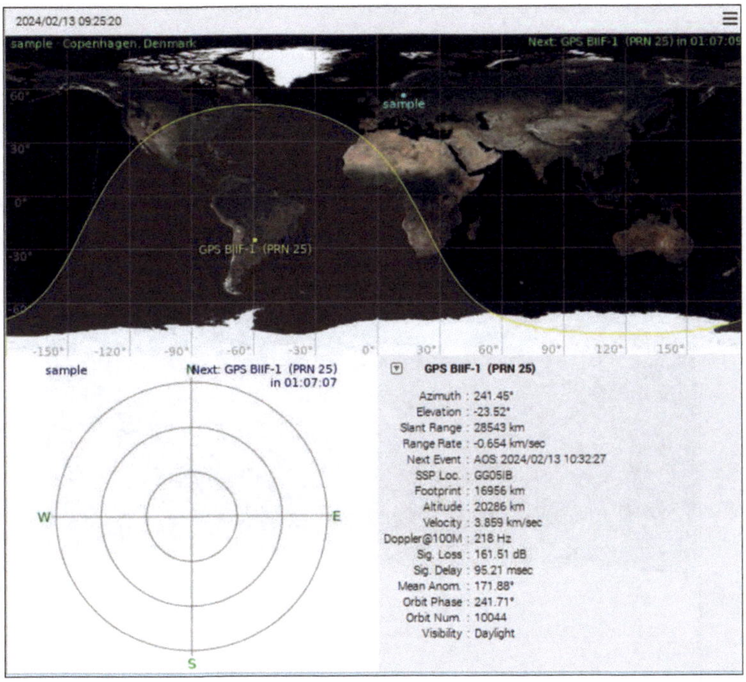

Figure 17.5: Map and the selected satellite.

The polar view (or radar view) shows the satellites on a polar plot, where the polar axis corresponds to the azimuth and the radial axis to the elevation. This graph can be helpful to give the position of the satellite in the sky.

Selecting a group of satellites

- The process is similar to above but move the required satellite names to the right hand pane. Figure 17.6 shows the selecting of 3 GPS satellites.
- The **Module Name** is set to **GPS3** in this example, and Figure 17.7 shows the map and satellite positions. You can double-click on a satellite to display data about it (Figure 17.8).

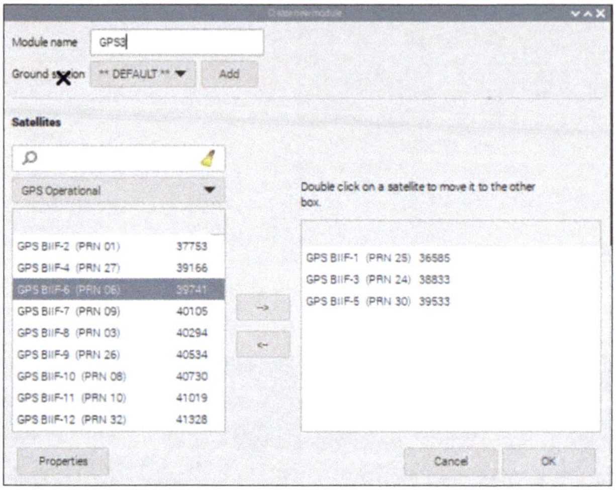

Figure 17.6: Select 3 GPS satellites.

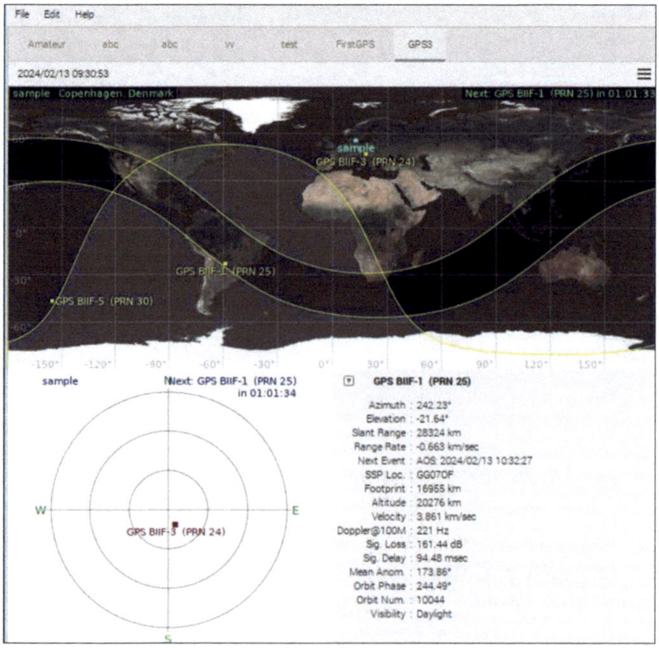

Figure 17.7 Map and selected satellites.

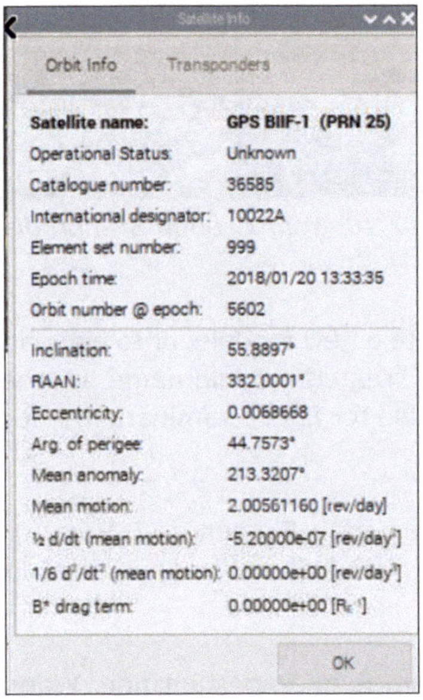

Figure 17.8: Double click on a satellite.

The Ground Station

The Ground Station details are below the Module name. this is the geographical location that is used as reference point in the calculations. The Ground Station should be defined accurately as it is used in the calculations. By default the Ground Station is set to a place in Denmark, called **sample**. You can add your Ground Station by clicking the **Add** button. A form will be displayed as in Figure 17.9. Enter your Ground Station details. For example, you can choose a location name, e.g. the name of your city and country. You will notice that your Ground Station will be displayed on the map (Figure 17.10).

Note: You should set your Ground station details before observing the satellites and their data.

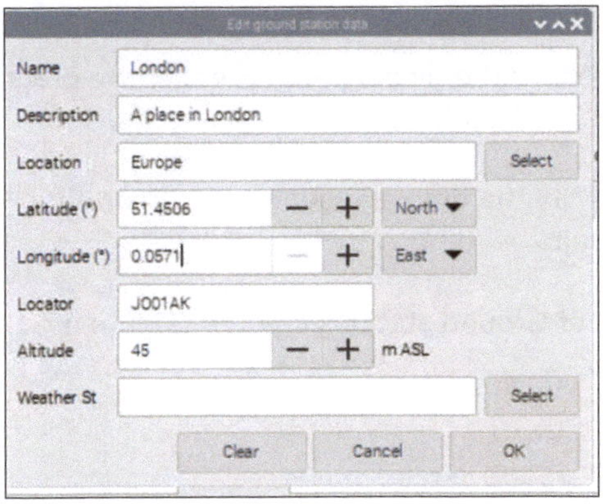

Figure 17.9: Ground Station details.

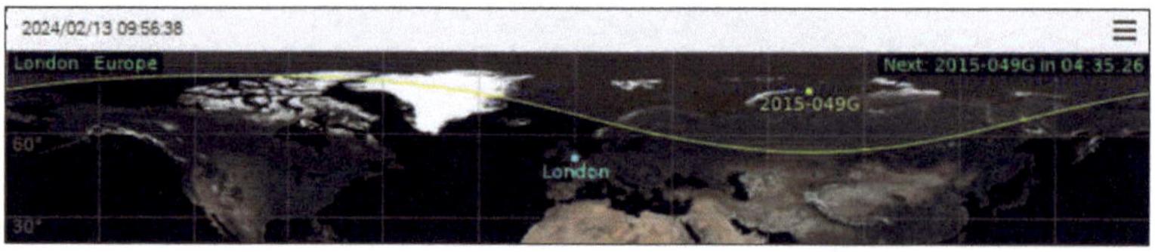

Figure 17.10: Your Ground Station displayed.

File menu

This menu enables you to create a new module, or to open an existing module, or to display the log. Select the previously created module name in order to open it. GPredict logs all run-time messages into a log file for later examination — for example, if there are errors.

Edit menu

GPredict needs up to date Keplerian Elements (TLE data) for the satellites. This menu updates the TLE data either from the network or from local files. It also updates the transponder data.

The Preferences sub-menu item is an important one. When opened, Figure 17.11 shows the display.

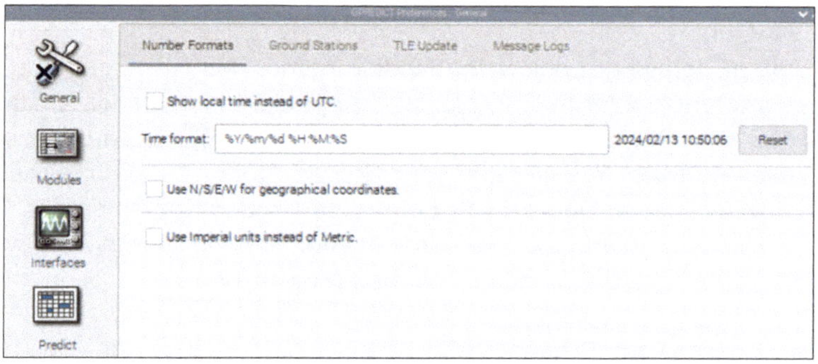

Figure 17.11: The Preferences menu item.

The menu items are:

General — This option defines the global parameters of the program, such as the Number Formats, Ground Stations, TLE Update, and Message Logs.

Number Formats — Specify the time and geographical formats, local time instead of UTC, imperial units or metric units.

Ground Stations — List of Ground stations created (Figure 17.12)

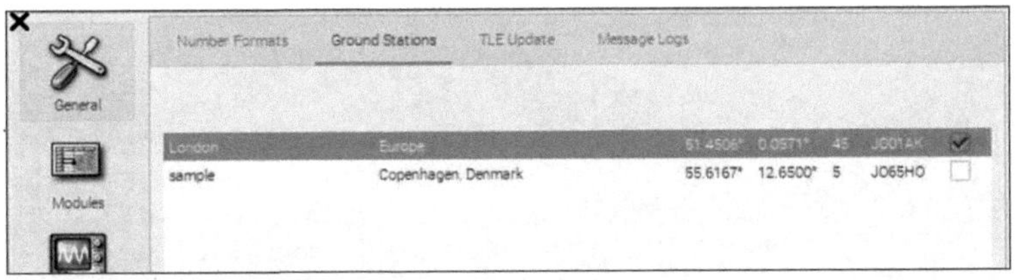

Figure 17.12: Ground Stations.

TLE Update — auto update, TLE sources, etc.

Message logs — debug level and deleting log files.

Modules

In this menu, you can select the Layout, Refresh Rates, List View, Map View, Polar View, and Single Sat View.

Layout — Select the display layout and window placements. Figure 17.13 shows the layout options.

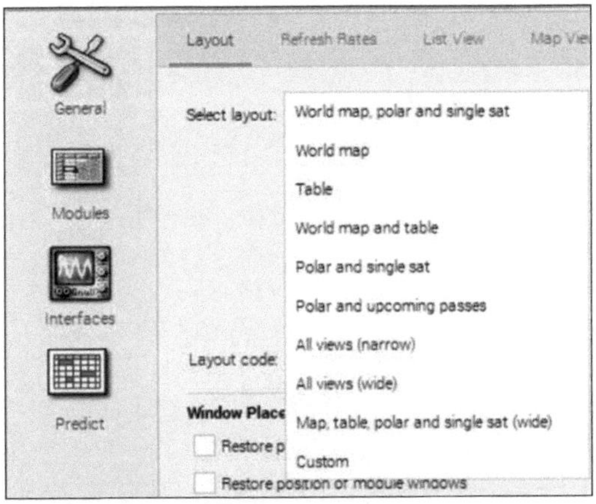

Figure 17.13: Layout options.

As an example, Figure 17.14 shows the display when the **All Views (Narrow)** is selected, where information is displayed below the map for the selected satellites.

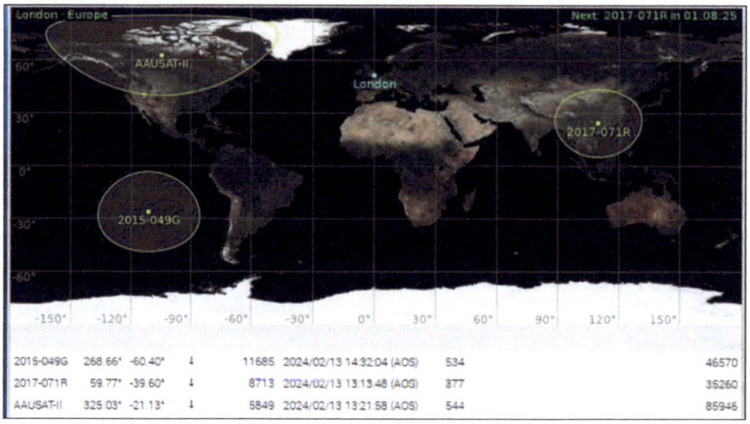

Figure 17.14: All View option.

Selecting option **Polar and upcoming passes** displays as in Figure 17.15.

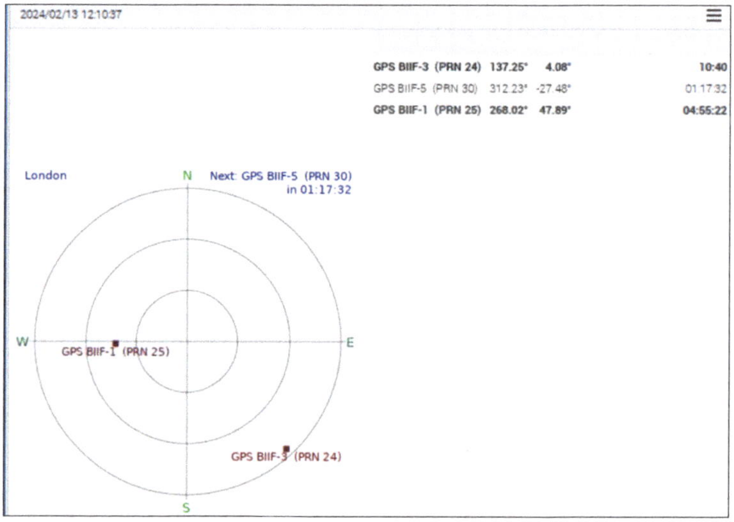

Figure 17.15 Polar and upcoming passes.

Refresh Rates — This menu option enables the refresh rates to be set (Figure 17.16)

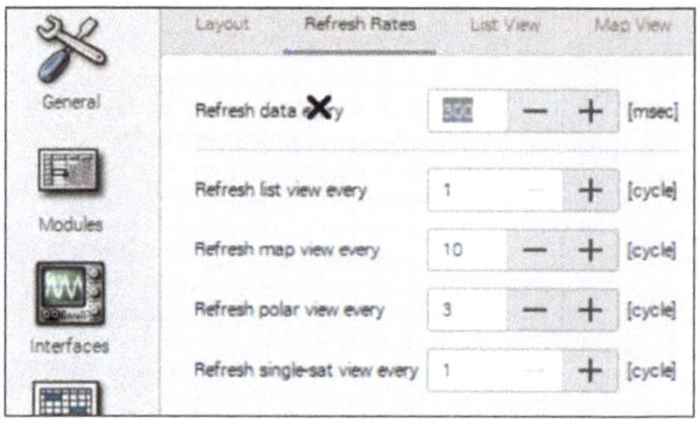

Figure 17.16: Set the refresh rates.

List View — Use this menu option to select the items to be displayed when the List Menu is selected. The defaults are ticked on the menu option.

Map View — Use this menu option to configure the map display, such as the colours etc.

Polar View — Use this menu option to configure the polar view display, such as the colors etc.

Single Sat View — Use this menu option to select the items to be displayed when a single satellite is selected. The defaults are ticked on the menu option.

Interfaces
Use this menu to configure the radios and rotators that you might be using in your station. The GPredict User Manual gives lots of information on how to use this menu.

Predict
Use this menu option to predict satellite passes in the future. The parameters and options on this page define how Gpredict predicts future passes for satellites. You can select Pass Conditions, Multiple Passes, Single Pass, and Sky at a Glance.

Pass Conditions — You can select various parameters such as the minimum elevation, number of passes, pass details, satellite visibility etc.

Multiple Passes — Use this menu option to select the items to be displayed when the Multiple Passes is selected. The defaults are ticked on the menu option

Single Pass — Use this menu option to select the items to be displayed when the Single Pass is selected. The defaults are ticked on the menu option

Sky at a Glance — Use this menu option to select the time within which the passes should occur, and also customize the satellite colours (Figure 17.17)

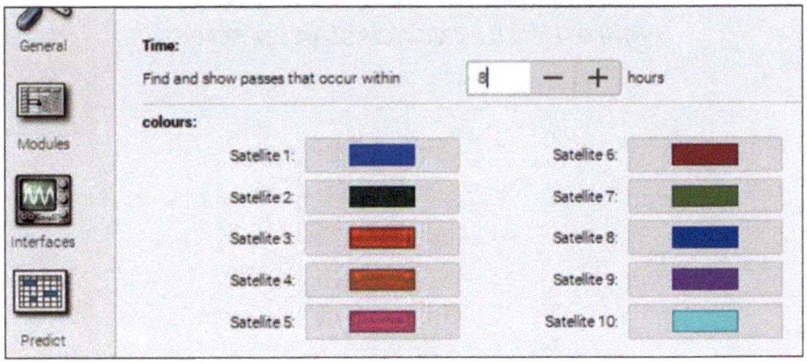

Figure 17.17 Sky at a Glance

17.5 TWCLOCK

This program (by WA0EIR) displays the local time and GMT in hundreds of major cities around the world. The program also has a timer, which can be set so that the station ID is output as audio Morse code at the specified time intervals. The sound output can be connected to the Audio In pin of your rig's accessory jack to have it transmitted automatically at the set intervals.

17.5.1 Installation on Raspberry Pi 5

The program can easily be installed from the Raspbian repository in desktop as in the previous installations in this chapter. Enter **teclock** to search box and wait until the program is installed.

17.5.2 Using the program

To start the program, open a terminal session in Desktop and enter:

pi@raspberrypi@~ $ **twclock**

Hold at the bottom right of the display to expand it as shown in Figure 17.18. Right click at the bottom of the twclock screen to display its options (Figure 17.19)

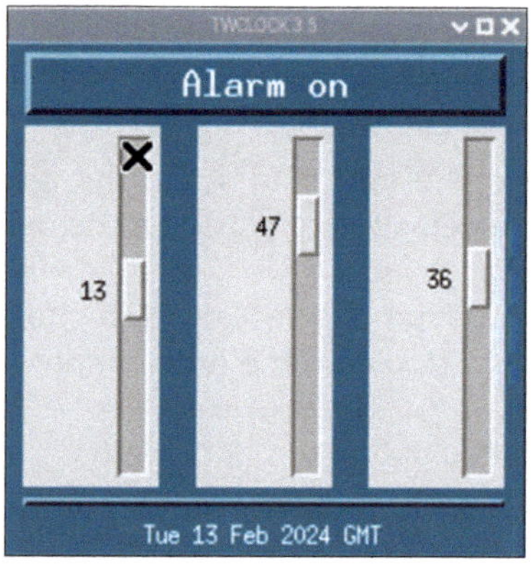

Figure 17.18: twclock startup screen.

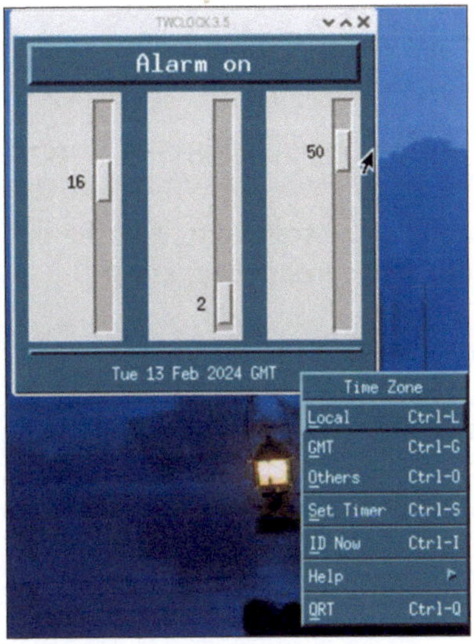

Figure 17.19: twclock options.

To see the world time, click the right button of your mouse and select **Others**. Select **Region** (e.g. Europe), select **City** (e.g. Turkey), click **OK**. You should see the time in Turkey displayed as shown in Figure 17.20.

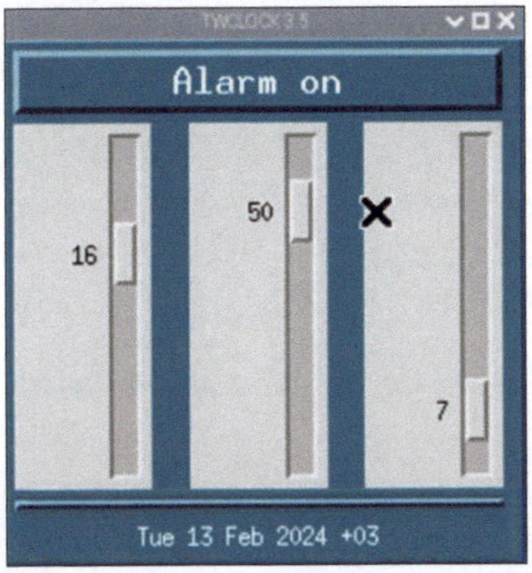

Figure 17.20: Displaying time in the selected city.

twclock is installed by default to the following directory:

/etc/X11/app-defaults

Before using the ID and the timer we have to specify our station ID (call sign) and the Morse code to be transmitted at the set timer intervals. This is done as follows:

- Use the nano editor and edit file:

pi@raspberrypi:~ $ **sudo nano /etc/X11/app-defaults/Twclock**

- Specify your station ID by editing the following statement (e.g. assuming the ID is **G7SCU**):

  Twclock.form.call_toggleB.labelString: **G7SCU**

- Specify the Morse code to be transmitted at the set time intervals (you can also change the CW speed, tone frequency, etc.):

  Twclock.cwStr: **de G7SCU**

- You may like to look at the other settings and change them if you wish.

- You should now re-start the twclock for the changes to take effect

Clicking ID Now in the main menu plays your station ID as an audible Morse code and you should hear it from your speaker. The time interval to play the station ID can be set by clicking Set Timer in the main menu (see Figure 17.21) and then entering the time in minutes and seconds, followed by OK. You should click CW ID to enable playing the CW. Setting Auto Reset will repeat the audio output at the specified intervals. The CW speed and the tone frequency can also be set from this menu.

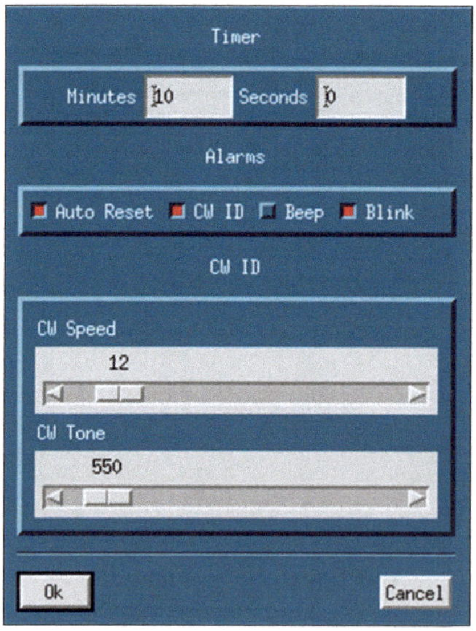

Figure 17.21: Set Timer selected.

A help option is provided in the menu for displaying information on various features of the twclock.

17.6 CQRLOG

CQRLOG is one of the popular amateur radio logging programs (by Petr, OK7AN and Martin, OK1RR), based on the MySQL database. The program provides radio control based on hamlib libraries, online callbook, internal QSL manager database support. CQRLOG is intended for daily general logging of HF, CW, and SSB contacts.

17.6.1 Installation on Raspberry Pi 5

Enter the following command to install the CQRLOG on Raspberry Pi 5:

pi@raspberrypi:~ $ **sudo apt-get install cqrlog**

17.6.2 Running the program

Run the program in Desktop by selecting **CQRLOG** from menu item **Hamradio** in the **Applications Menu**. Alternatively, start a terminal session in Desktop and type **cqrlog**.

When the program is run the first time, it creates the MySQL database and this may take several minutes. Click Open Log to open the log as shown in Figure 17.22.

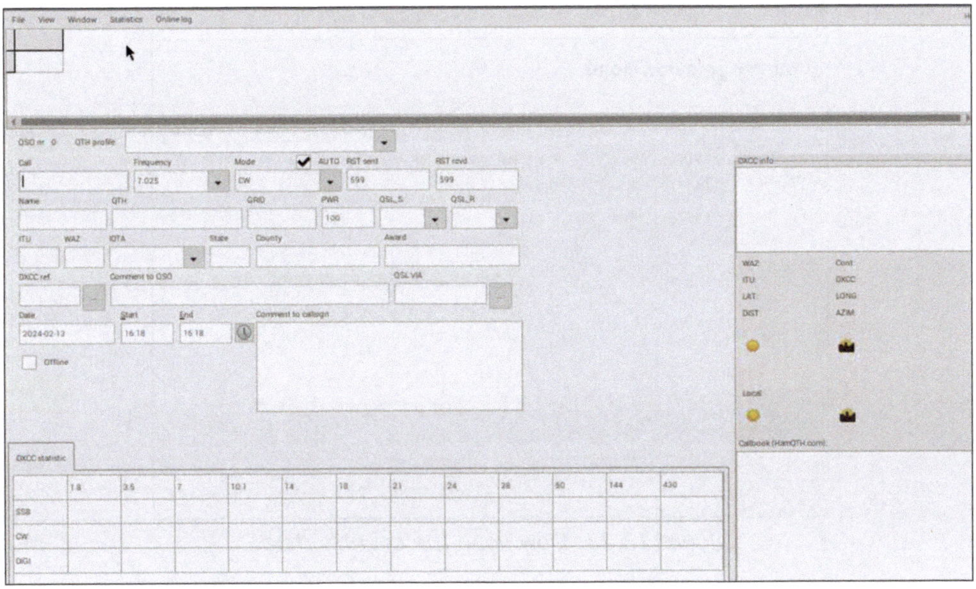

Figure 17.22: Main screen of CQRLOG.

Detailed information on using the program is available at the following web site:

https://www.cqrlog.com/help/index.html
http://www.cqrlog.com

17.7 Klog

This program (by Jaime Robles, EA4TV) has been developed to replace paper logbook. The program is easy to use and is available in many platforms (e.g. Windows, MAC etc.) and in many languages. Klog provides QSo management, QSL management, DXCC management, club log integration, and much more.

17.7.1 Installation on Raspberry Pi 5

The program should be installed as described in Desktop using the **Applications Menu** but do enter **klog** for the program name. The program is listed as **Multiplatfoem ham radio logging program**.

17.7.2 Using the program

To start te program, open a terminal session in Desktop and enter:

pi@raspberrypi@~ $ **klog**

When the program is run for the first time, you will be asked to accept a license condition and click to download country data.

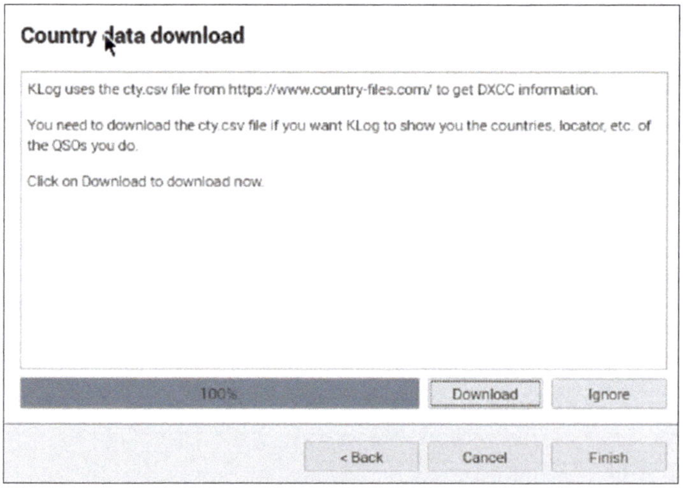

Figure 17.23: Download the country data.

Then, a **Config Dialog** screen will be displayed to enter configuration data e.g. callsign, personal data, station data, band details, modes of operation etc. Figure 17.24 shows the main screen of klog. The program is very comprehensive and includes many menus and options. At top left we have the entry box, top right the output box, and the at the bottom part we have the Log, DX-Cluster, and DXCC. Perhaps the best way to learn to use this program is to download it and start playing with it.

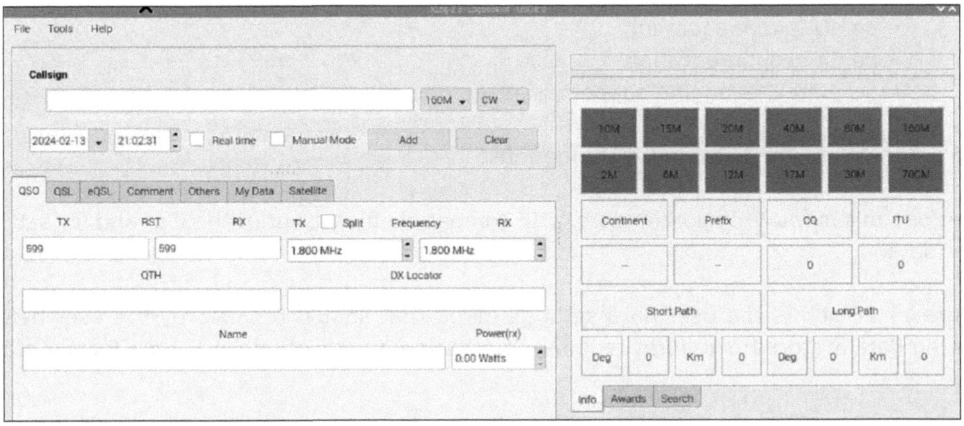

Figure 17.24: Klog main screen.

As an example, Figure 17.25 shows a log assuming the contact callsign is ABCD and you made a contact on the 20th of January 2023 at 10:00:00 with voice clarity 5-9, on 14 MHz, to Mr. A. Jones and running 5 watts of RF power. Click **ADD** and you should see the data added to the log at the bottom part of the display.

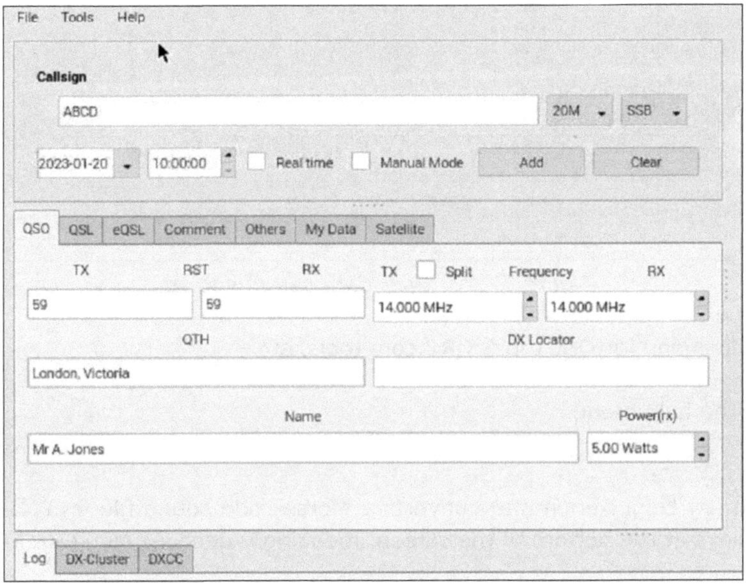

Figure 17.25: Example log.

Using klog, the operator can:

- add/edit/search/remove QSOs
- manage QSLs
- import data from formats such as ADIF, TLF, Cabrillo
- support DXCC and WAZ local awards

- do eQSL management
- do ClubLog integration
- get Satellite logging support

The program has the following menu options:

File: use this menu to import/export ADIF compatible files, to print the log, and for setting user data.

Figure 17.26 shows the user data settings menu that should be filled by the user before any log data is stored. The form requires the callsign, bands of interest, time format etc to be entered.

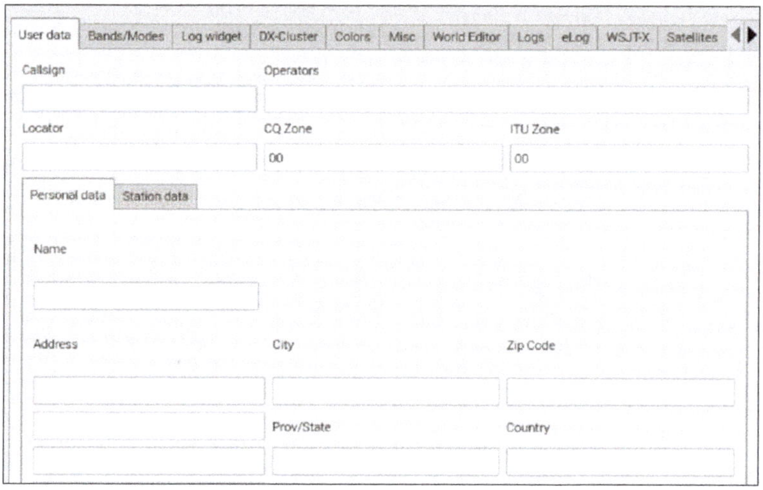

Figure 17.26: User data settings form.

Tools: use this menu for QSL tools, QRZ.com tools, etc.

Help: this is the help menu.

17.8 Morse2Ascii

This program (by Luigi Auriemma) converts a Morse code sound file in .WAV format into text and displays at the bottom of the screen, meaning it decodes Morse code sound files.

17.8.1 Installation on Raspberry Pi 5

The program should be installed as described in Desktop using the **Applications Menu** but do enter **morse2ascii** for the program name. The program is listed as **tool for decoding the morse codes from a PCM WAV file**.

17.8.2 Using the program

Enter the command **morse2ascii** followed by the WAV filename. The program will display the decoded code in text at the bottom of the display. An example run of the program is shown in Figure 17.27. Note that file morse.wav was created by the authors which include the sound for morse text **this is a test morse code**.

```
pi@raspberrypi:~ $ morse2ascii morse.wav

MORSE2ASCII 0.2
by Luigi Auriemma
e-mail: aluigi@autistici.org
web:    aluigi.org

- open morse.wav
  wave size      87280
  format tag     1
  channels:      1
  samples/sec:   8000
  avg/bytes/sec: 8000
  block align:   1
  bits:          8
  samples:       87280
  bias adjust:   0
  volume peaks:  -31744 31744
  normalize:     1023
  resampling to  8000hz

- decoded morse data:
this  is  a  test  morse  code
pi@raspberrypi:~ $ ▮
```

Figure 17.27 Running the program

Note that you can use the following website to convert text into a Morse code WAV sound file for testing:

https://textcleaner.net/morse-code-translator/

Morse2Ascii program accepts the options shown in Figure 17.28 (taken directly from the program by just entering **morse2ascii**:

```
Options:
-a        abbreviations and prosigns parsing
-q        q-codes parsing

-r F C B  consider the file as raw headerless PCM data, you must specify the
          Frequency, Channels and Bits like -r 44100 2 16
-o        disable the automatic optimizations: DC bias adjust and normalize.
          use this option only if your file is already clean and normalized
-m        morse notation output (like ...___... instead of SOS), debug
-w FILE   debug option for dumping the handled samples from the memory to FILE
-d        debug info (WAV input only)

Note: the input file can be a PCM WAV audio file or also a text file which uses
      the dotlinespace notation (._ or .-), dit-dah, binary and others
      use - as file for reading a text stream from stdin
```

Figure 17.28: Program options.

17.9 PyQSO

This is another amateur radio operator logging program (by Christian Thomas Jacobs).

17.9.1 Installation on Raspberry Pi 5

The program should be installed as described earlier in Desktop, using the **Applications Menu**, but do enter **pyqso** for the program name. The program is listed as **logging tool for amateur radio operators**

17.9.2 Using the program

Run the program from Desktop by clicking **Application Menu → Hamradio → pyQSO** or open a terminal session in Desktop and enter **pyqso**.

Figure 17.29 shows the startup screen of the program. The menu options are Logbook, Records, View, and Help.

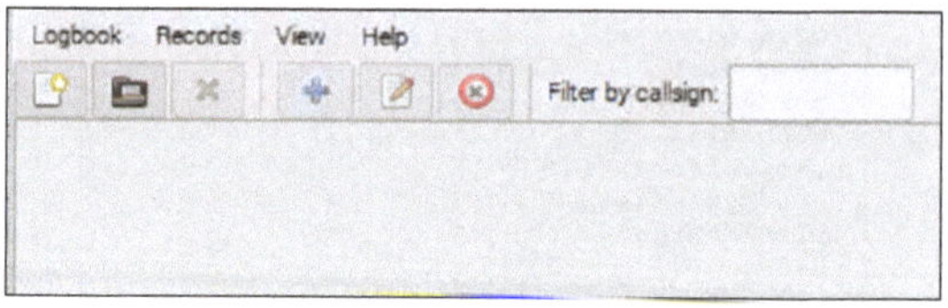

Figure 17.29: pyQSO startup screen.

Logbook: use this option to create a new logbook, to open an existing logbook, import and export ADIF formatted files, print a logboog, and preferences. In the Preferences sub-menu you can show yearly statistics, set the visible fields, set the modes, TX power, callsign lookup, world map, etc.

Records: you can add/edit/delete records display record count.

To create a new log, click **Logbook → New log** and enter a log name. Initially the screen shown in Figure 17.30 is created. Click **Records** to add a new record to logbook. Click **+** for a logname. Click **Records → Add record** to Add record to the logbook. Click **OK** at the bottom of the form to save it. An example is shown in Figure 17.31.

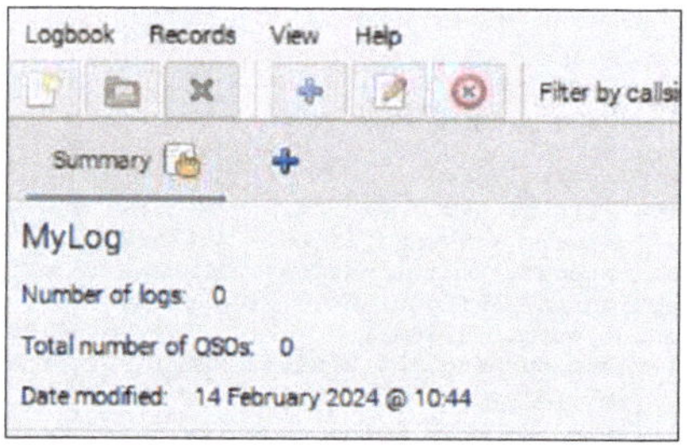

Figure 17.30: Initial screen.

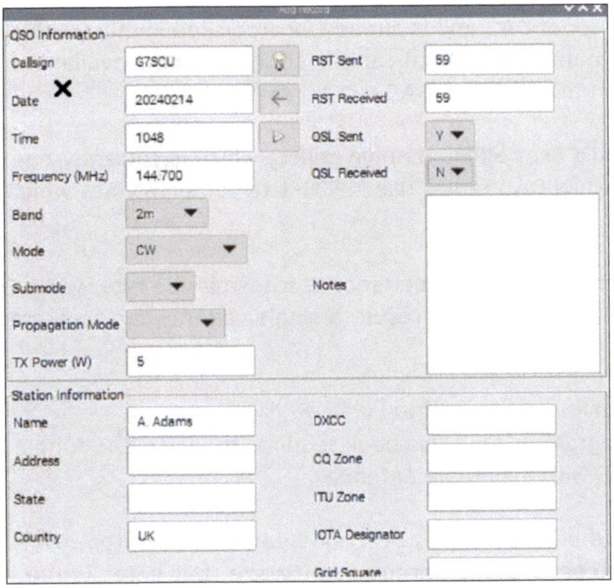

Figure 17.31: A new log.

Figure 17.32 shows the main screen after the log is added. Clicking **Records → Record Count** will display '1' as there is only one record in the logbook.

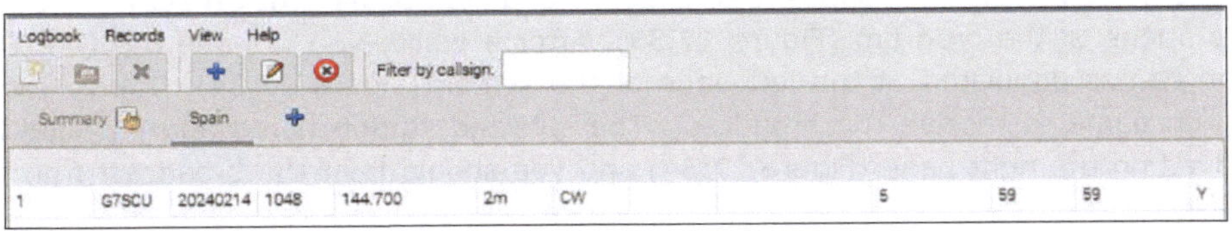

Figure 17.32: Main menu after adding a log.

17.10 Welle.io (DAB/DAB+ Radio)

DAB stands for **D**igital **A**udio **B**roadcasting, where sound is transmitted digitally. Most readers are familiar with the popular classic FM radios. DAB radios actually are successors of classic FM radios. DAB radios have the great advantages that they are of high quality even at remote locations, provided you are within their reception ranges.

FM radios are analog where radio waves are transmitted as analog signals. DAB radios on the other hand are digital transmissions and they can be received without any static noise caused by radio waves. Digital radios usually have automatic channel search functions, not requiring the manual selection of stations. It is expected that in 2030s all FM radios will be replaced by DAB radios.

There is also differences between DAB radios and internet based radios. Internet radios require reliable stable internet connections and they stream the programmes online, while the DAB radios receive content digitally without streaming. Listening to DAB radio is like listening to a CD.

DAB+ is the successor of DAB and is already incorporated in many vehicles in the European maket. Increasing number of digital radio stations are now available in DAB+. The main differences between the DAB and DAB+ radio are:

- DAB+ radio uses MPEG-4 audio codec, which is currently one of the best codecs, while DAB radios use MPEG-1 to compress files which is inferior to MPEG-4.

- DAB+ works with 80 kB per second transmission rate, while DAB works with 128 kB per second. As a result, a single channel can accommodate multiple stations.

17.10.1 Installation on Raspberry Pi 5

The DAB/DAB+ program used in this book requires the RTL-SDR dongle to be connected to your Raspberry Pi 5 with a suitable antenna.

The program should be installed as described earlier in Desktop, using the **Applications Menu → Preferences → Add/Remove software**, but enter **welle.io** for the program name. The program is listed as **DAB/DAB+ Software Radio**

17.10.2 Running the program

Open a Terminal session in Desktop and enter the command: **welle-io** to start the program. Click the three vertical lines opposite All Stations and click **Start station scan** to scan and load stations to the program (Figure 17.33). After a while, you will see the DAB/DAB+ station names populated at the left pane of the screen (Figure 17.34). Double-click on a station name to receive the broadcast. The selected station name and its details are displayed on the right pane (Figure 17.34) and you should hear the broadcast from your speaker.

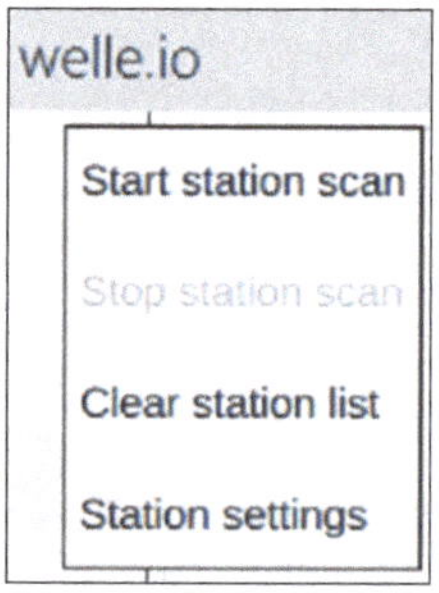

Figure 17.33: Scan for stations.

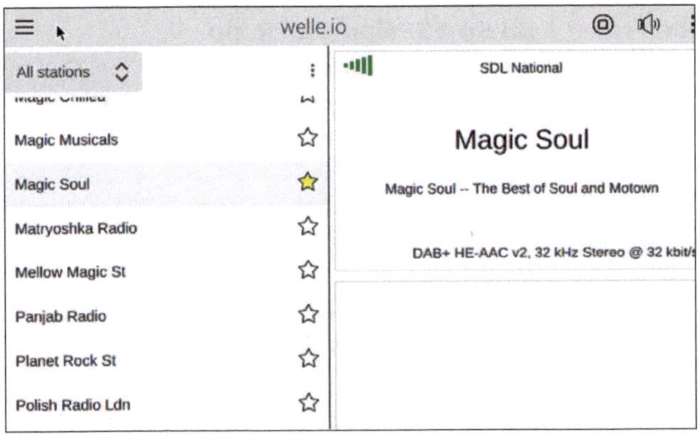

Figure 17.34: DAB/DAB+ Station names.

You have the option to clear all stations, as well as to set a station to play when the program is started (in **Station Settings**)

17.11 Ham Clock

The **Ham Clock** (by Karl-Heinz, DL1GKK) is a detailed dynamic display (see: https://dl1gkk.com/ham-clock-raspberry-pi/) which is updated in real-time and displays the following interesting and useful items (only some items are listed here):

- your callsign
- lLocal (or UTC) date and time
- sunlight position and illumination
- analog clock
- satellite information
- XRay or Kp indx or VOACAP
- sunspots or solar flux
- solar images
- NCDXF beacons
- orbital predictions to show when your favorite ham satellites are going to be overhead.

17.11.1 Installation on Raspberry Pi 5

The steps to install the software are as follows:

- pi@raspberrypi:~ $ **sudo apt-get update**
- pi@raspberrypi:~ $ **sudo apt-get upgrade**
- pi@raspberrypi:~ $ **sudo apt-get dist-upgrade**
- pi@raspberrypi:~ $ **sudo reboot**

After the reboot, enter the following commands:

- pi@raspberrypi:~ $ **curl -o ESPHamClock.zip http://www.clearskyinstitute.com/ham/HamClock/ESPHamClock.zip**

- pi@raspberrypi:~ $ **unzip ESPHamClock.zip**
- pi@raspberrypi:~ $ **cd ESPHamClock**
- pi@raspberrypi:~/ESPHamClock $ **make -j 4 hamclock**

17.11.2 Running the program

- Enter the following command to run the program in Desktop after opening a terminal session:

  pi@raspberrypi:~ $ **cd ESPHamClock**
  pi@raspberrypi:~/ESPHamClock $ **./hamclock**

When the program runs you should do some configuration (see the User Guide at web site: https://www.clearskyinstitute.com/ham/HamClock/), where the program asks 5 pages of configuration, including your callsign, your latitude and longitude etc. Click DONE at the bottom of the configuration menu when completed.

Figure 17.35, Figure 17.36, and Figure 17.37 show two displays from the program. Notice that your location is marked on the map. Current date and time are displayed at the top left corner of the screen.

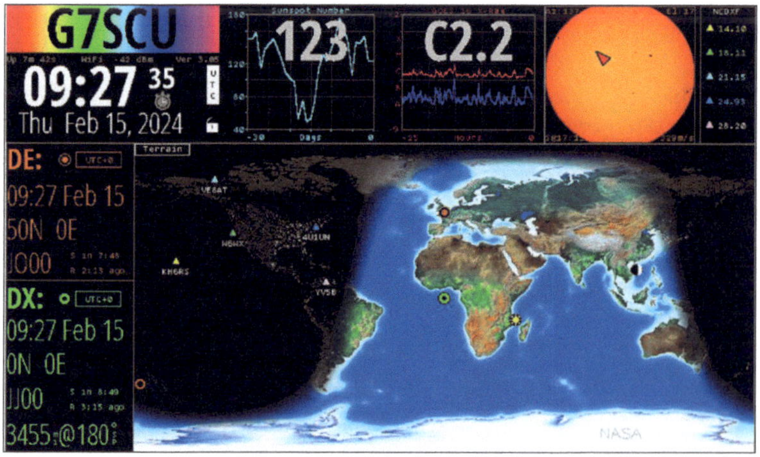

Figure 17.35: Display from the program.

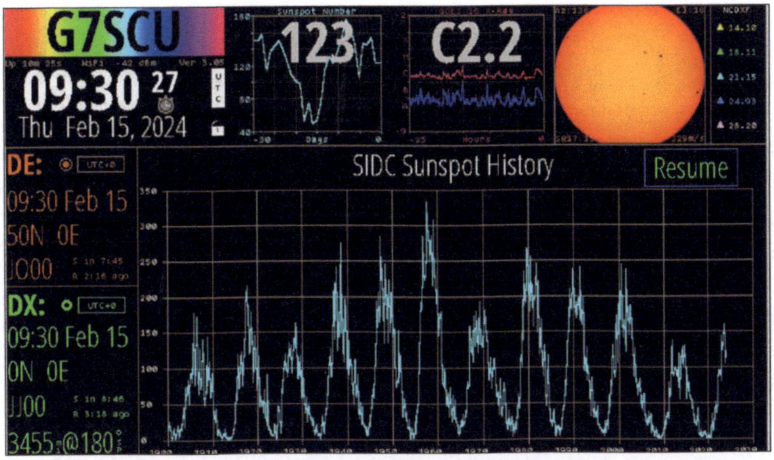

Figure 17.36: Another display from the program.

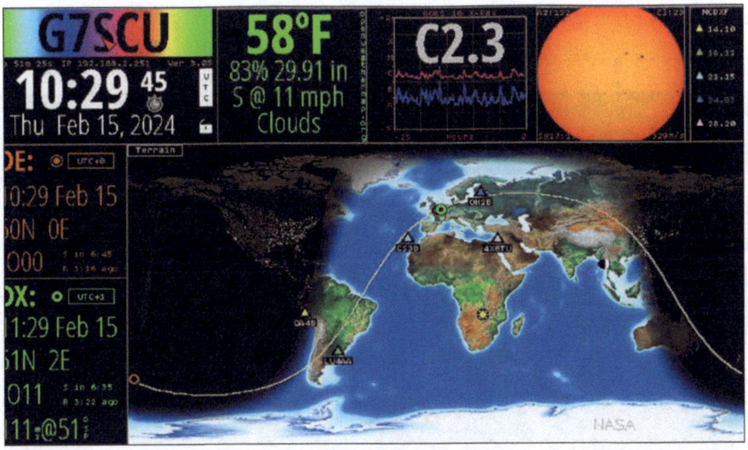

Figure 17.37: Another display from the program.

A Desktop shortcut can be created to run the program easily by clicking on this shortcut. The steps are (see the User Guide):

- Right click in Desktop to create an empty file on the Desktop named **HamClock.desktop**

- Edit the file and add the following statements

```
]Desktop Entry]
Name=HamClock
Comment=Open HamClock
Exec=/home/pi/ESPHamClock/hamclock
Type=Application
Encoding=UTF-8
```

```
Terminal=false
Categories=None;
```

17.11.3 Accessing from a web browser

Ham Clock can be accessed from a web browser (e.g. Firefox, Chrome etc). The steps are:

- Get the IP address of your Raspberry Pi 5.
- Start the Ham Clock program on your Raspberry Pi 5.
- Enter the following address on your web browser.

http://192.168.1.251:8081/live.html

Where 192.168.1.251 is the IP address of Author's Raspberry Pi 5. Figure 17.38 shows the Ham Clock displayed on a web browser

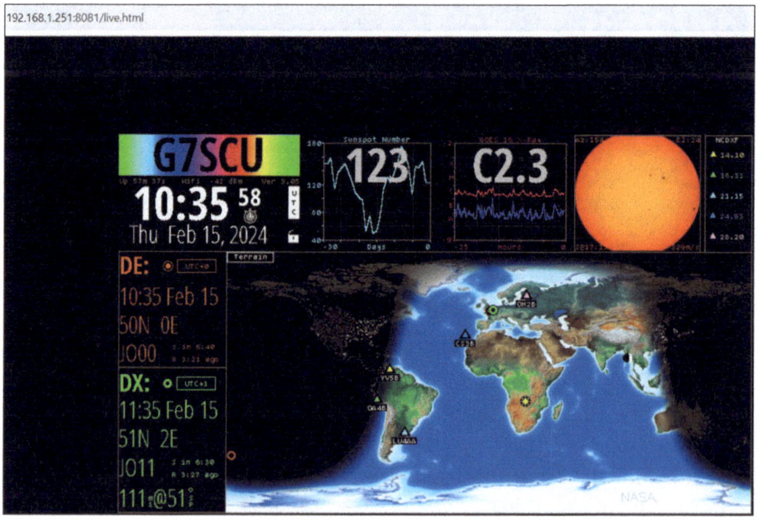

Figure 17.38: Displaying on a web browser.

Ham Clock also includes an alarm clock with a countdown timer. Click on the clock next to the time display to display the alarm clock (Figure 17.39). Enter the required countdown time and click Run to start or Stop to stop the timer.

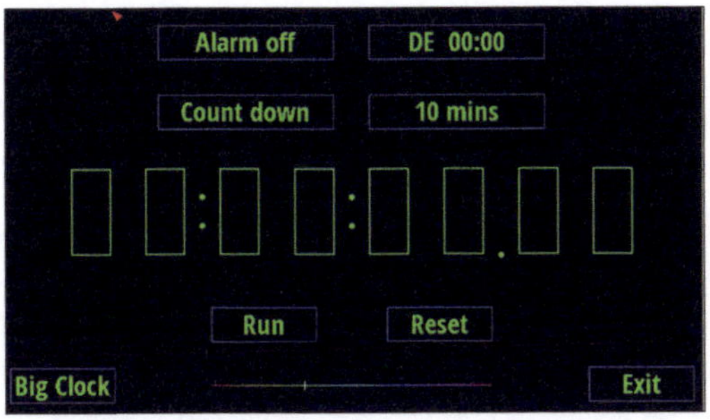

Figure 17.39: Alarm clock display.

17.12 Chirp

CHIRP is a popular tool for programming amateur radios. CHIRP is available on several platforms including the Raspberry Pi 5. The program supports a large number of manufacturers and models. It is especially popular for programming cheap Chinese radios such as Baofengs Wouxun etc. The CHIRP website given below provides detailed information on the supported models and how to use the software:

https://chirp.danplanet.com/projects/chirp/wiki/Home
also: https://chirp.danplanet.com/projects/chirp/wiki/Beginners_Guide

17.12.1 Installation on Raspberry Pi 5

You can install the program from the Desktop **Applications Menu → Preferences →Add/ Remove software**. Type **chirp** in the search box and press Enter. The program is shown with the name: **Configuration tool for amateur radios**. Click to select the program and click **Apply** to install it.

17.12.2 Running the program

The program can be run from the Desktop by clicking **Application Menu → Hamradio** and then **CHIRP**.

Figure 17.40 shows the startup screen when the program runs. Click **Radio** followed by **Download** to set your serial port and to select a manufacturer and a model. Note that the Raspberry Pi 5 serial port is named **/dev/tyAMA10** as shown in Figure 17.41. You should connect your radio to Raspberry Pi 5 using either serial connection or through the USB port depending on your radio and the type of cable. Depending on the type of radio you have, you will either see a progress display bar, or the program will jump to the memory editor when the download is complete.

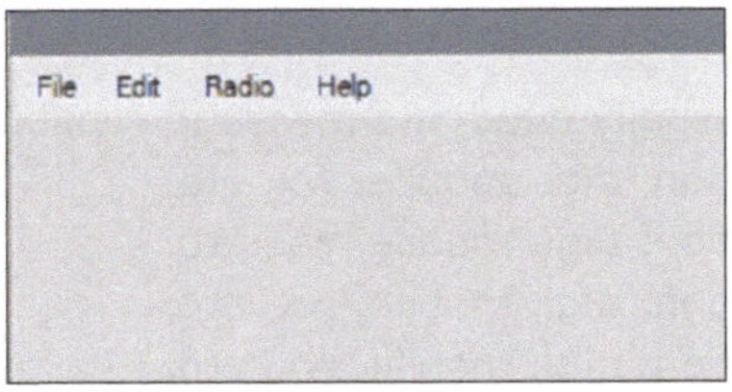

Figure 17.40: CHIRP startup screen.

Figure 17.41: Selecting the serial port.

Figure 17.42 shows the serial port (UART) on Raspberry Pi 5 board and the serial cable compatible with this port.

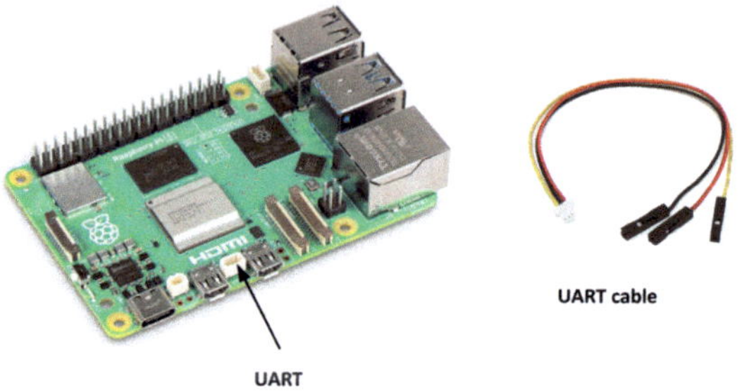

Figure 17.42: Raspberry Pi 5 serial port and cable.

17.13 Xastir

Xastir is an APRS program that provides mapping, tracking, messaging, weather, weather alerts, and search and rescue operations over radio. APRS is a digital communication system for real-time exchange of digital information. APRS (Automatic Packet Reporting System) is digital communication system for real-time exchange of digital information. It allows broadcast and reception of APRS traffic via an Internet connection or via a connected radio (e.g., to receive APRS stations in your area). APRS was originally developed by Bob Bruninga (WB4APR). The name XASTIR is an acronym for **X** **A**mateur **S**tation **T**racking and **I**nformation **R**eporting.

You need to provide your callsign to identify yourself and get a password as explained later in this section. You are encouraged to use the standard suffixes to differentiate the type of station you have (e.g., mobile, on boat, aircraft, etc). The map shows your suffix as an icon. For example, if your call sign is G5SUT and you are using your station on an aircraft, you would see G5SUT-11 on the map and an aircraft icon will be displayed. A list of the international suffixes are:

-0 Your primary station usually fixed and message capable
-1 generic additional station, digi, mobile, wx, etc.
-2 generic additional station, digi, mobile, wx, etc.
-3 generic additional station, digi, mobile, wx, etc.
-4 generic additional station, digi, mobile, wx, etc.
-5 Other networks (Dstar, Iphones, Androids, Blackberry's etc).

-6 Special activity, Satellite ops, camping or 6 meters, etc.
-7 walkie talkies, HT's or other human portable.
-8 boats, sailboats, RV's or second main mobile.
-9 Primary Mobile (usually message capable).
-10 internet, Igates, echolink, winlink, AVRS, APRN, etc.
-11 balloons, aircraft, spacecraft, etc.
-12 APRStt, DTMF, RFID, devices, one-way trackers*, etc.
-13 Weather stations.
-14 Truckers or generally full-time drivers.
-15 generic additional station, digi, mobile, wx, etc.

17.13.1 Installation on Raspberry Pi 5

You can install the program from the Desktop **Applications Menu → Preferences → Add/Remove software**. Type **xastir** in the search box and press Enter. Click to select the program and click **Apply** to install it.

17.13.2 Running the program and configuring for internet APRS

Before running the program, you should get a password by entering the command:

```
callpass YOURCALL
```

Where YOURCALL is your callsign. A numeric password will be displayed on the screen. Keep this in a safe place as you will need it while running xastir.

Open a terminal session in Desktop and enter the following command:

pi@raspberrypi:~ $ **sudo xastir**

When the program starts for the first time, you should see the **Configure Station** menu (Figure 17.43) where you should enter your callsign and the LAT/LONG of your station etc. and click **OK**. You should then see the main screen as shown in Figure 17.44. Click the **IN** button to zoom in to the map and use the four arrow keys next to **OUT** button to navigate through the map. Figure 17.45 shows the map zoomed in to the UK.

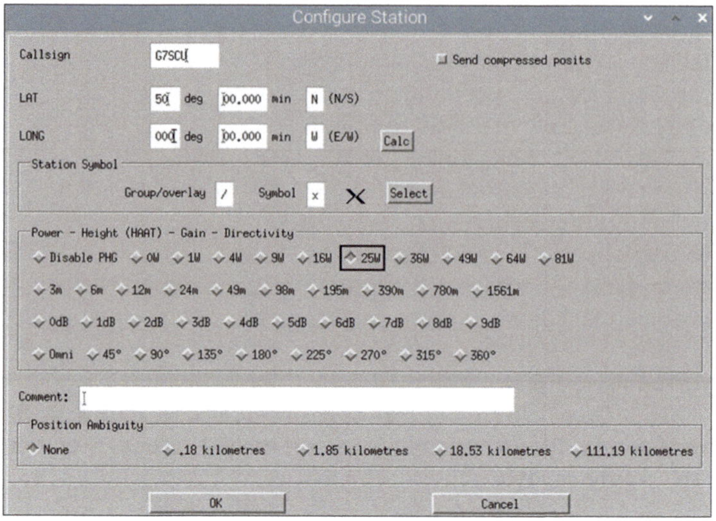

Figure 17.43: Configure Station menu.

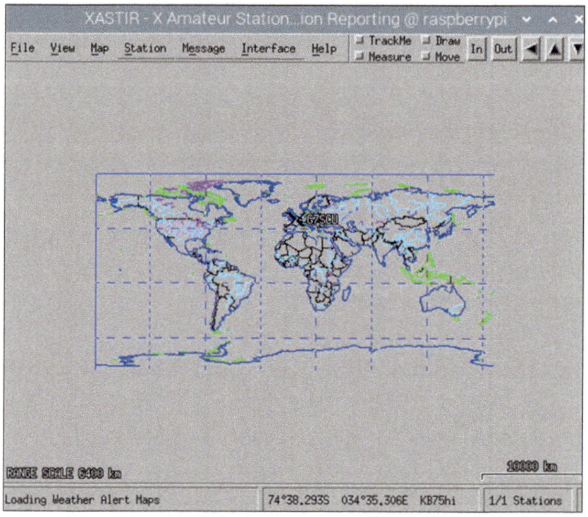

Figure 17.44: Main menu.

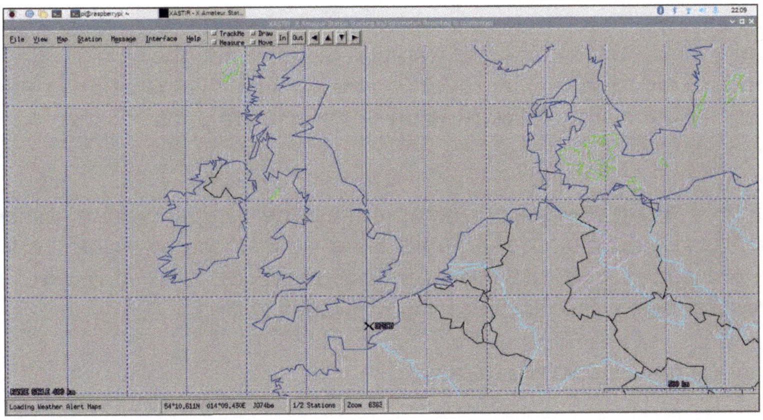

Figure 17.45: Map zoomed in.

Configuring APRS Internet server

Click **Interface → Interface Control → Add → Internet Server → Add**

You will have to enter an APRS internet server address and port name. The authors used the following server and port:

london.aprs2.net:14580

Where **14580** is the port number, and **london.aprs2.net** is the Internet APRS server name. Enter these details on the form. You will also be asked to enter your **Pass-code** which is the password you were given in response to **callpass**. Click **OK** to accept. The status of this server van be viewed at web address: http://london.aprs2.net:14501/. You can add more servers if you wish.

APRS filters

You can use filters for your server connection. For example, you can specify to show N stations within M kilometres away from your station, show all stations, search for a given station, search for specific prefixes etc.

Map menu

Click **Map → Map Chooser** to select a map type, etc. There re many Map options that you can choose from.

View menu

You can view a list of all stations by clicking **View → All Stations**. Click **View → Mobile Stations** to display a list of mobile stations. Click **View → Weather Stations** to display a list of the weather stations. Click **View → Incoming Data** to display the incoming packet data etc.

Station menu
Click **Station → Find Station** to locate a station by specifying the callsign. Click **Station → Track Station** to track a station by specifying the callsign. Click **Station → Filter Display** to select filters for the display. Click **Station → Filter Data** to filter data, etc.

Message
Click **Message → Send Message** to send a message by specifying a station's callsign. Enter the message as text and click **Send Now** to send the message. Click **Messsage → Pending Messages** to display any pending messages. Click **Message → Clear All Outgoing Messages** to clear any outgoing messages etc.

For further details on using **xastir**, head to:

https://xastir.org/index.php/Main_Page
https://www.george-smart.co.uk/aprs/ui_view_aprs_is_settings/

17.14 QSSTV

SSTV is a means of sharing images among amateurs. It was introduced by Copthorne Macdonald back in 1957-58, where he used an electrostatic monitor and a Vidicon tube. He managed to transmit black and white 120 lines and about 120 pixels per line still picture using a 3 kHz wide telephone channel.

Using a receiver capable of demodulating single-sideband modulation, SSTV transmissions can be heard on the following frequencies:

160 m	-	1.890 MHz, LSB, analog
		1.924 MHz, LSB, analog
80 m	-	3.845 MHz (3.73 MHz in Europe), LSB, analog
40 m	-	7.171 MHz (7.165 MHz in Europe), LSB, analog
20 m	-	14.230 MHz, USB, analog
10 m	-	28.680 MHz, USB, analog
6 m	-	50.680 MHz, FM, analog
		50.950 MHz, USB, analog
2 m		144.500, FM, analog
		145.550 MHz, USB, analog
		145.800 MHz, FM, analog
70 cms		430.950 MHz, USB, analog

SSTV uses the audio band to send image. In a typical application, the computer (e.g. Raspberry Pi 5) audio adapter is connected to the audio interface of the rig. Then a program (like qsstv) runs on the computer to receive the images. There are several SSTV frequencies in the amateur band, such as in 2 meters (144.550 MHz, 145.500 MHz, 145.600 MHz), on 20 meters (14.230 MHz and 14.233 MHz), and others. This is the Slow Scan TV application (by Johan Maes, ON4QZ) program QSSTV.

17.14.1 Installation on Raspberry Pi 5

You can install the program from the Desktop **Applications Menu → Preferences → Add/Remove software**. Type **qsstv** in the search box and press Enter. Click to select the program and click **Apply** to install it. The version installed by the authors was V9.5.8.

17.14.2 Running the program

To start the program, open a terminal session in Desktop and enter the following command:

pi@raspberrypi:~ $ **qsstv**

Alternatively, you can click Applications Menu in Desktop and then click QSSTV.

Figure 17.46 shows the qsstv startup menu. When the program is run the first time it should be configured. Click **Options** menu and then **Configuration** and enter the operator details, Directories, GUI, audio details, CW speed etc (see Figure 17.47 for operator details). Click **OK** when finished. Figure 17.48 shows the audio configuration.

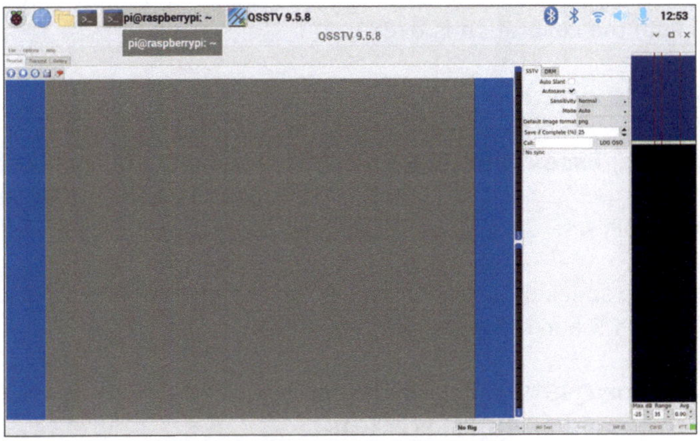

Figure 17.46: qsstv startup menu.

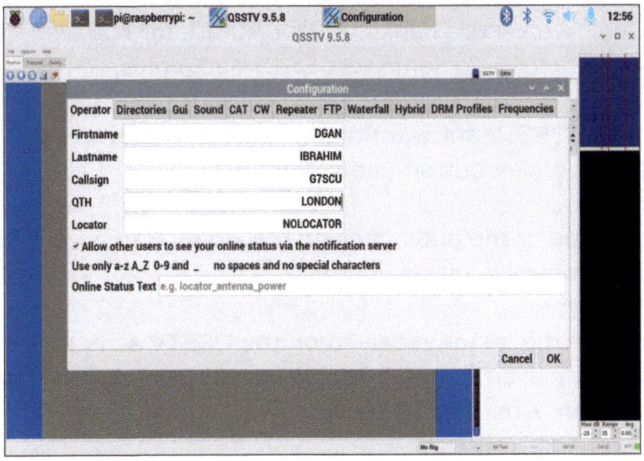

Figure 17.47: Configure operator details.

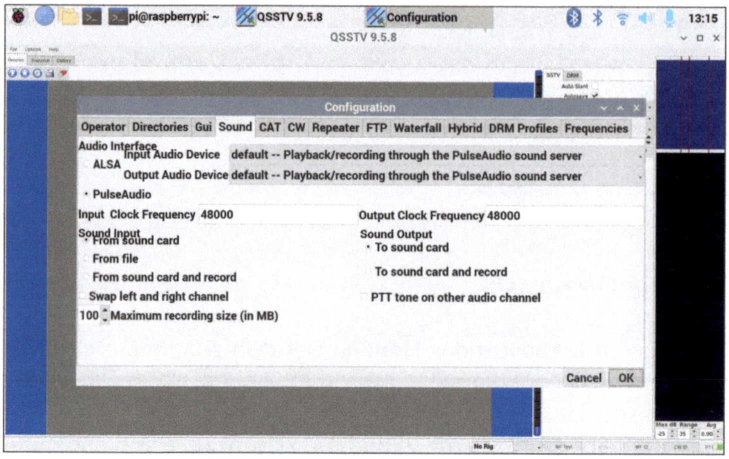

Figure 17.48: Audio configuration.

Click **Options** and then **Calibrate** to calibrate your sound card. You may have to wait for several minutes until the calibration is over.

17.14.3 Testing the program

There are many SSTV audio files available on the Internet that can be used to test the program. For example, **Essex Ham (UK)** provides 4 audio CCTV signals on their website: https://www.essexham.co.uk/sstv-the-basics. The steps to use one of these signals to test the software are as follows:

- Connect the speaker output of your PC to the microphone input of your Raspberry Pi 5 audio adapter (e.g. UGREEN).
- Configure the QSSTV software to receive audio data from the audio adapter USB port (Figure 17.48) and click **OK**.
- Open the **Essex Ham** web site where the audio files are located. Four files are given on their web site. Thanks to Paul M0CNL for suggesting that it would be handy to have access to some test SSTV audio files.
- Make sure the QSSTV software menu is set in **Receive** mode and click the start button (the blue play button under **Receive**).
- Click to play one of the audio files on the Essex Ham web site, making sure that your volume control is up on your PC.
- You should see the image received on the QSSTV screen as shown in Figure 17.49. The file played in this example was named: **Essex Ham 01 Scottie2 MP3**, it lasts for 1 minute 12 seconds.

Figure 17.49: Image built on QSSTV screen.

There are several other test audio files on this web page: https://www.sigidwiki.com/wiki/Slow-Scan_Television_(SSTV). Figure 17.50 shows the example file Scottie1 taken from this website.

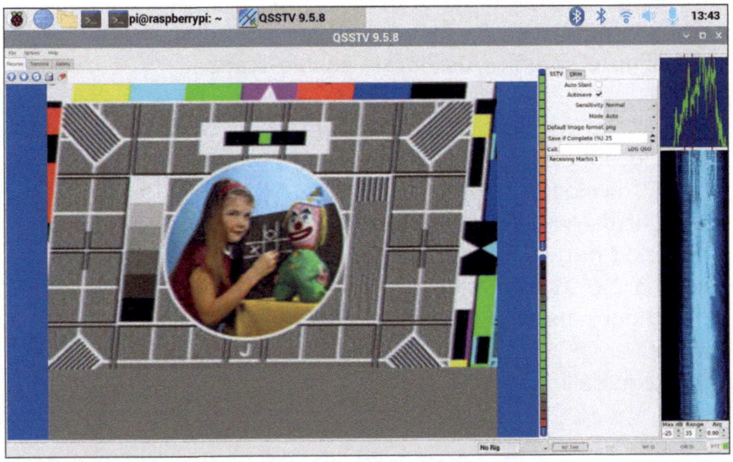

Figure 17.50: Example image built on QSSTV screen.

17.15 FLRIG

FLRIG is a transceiver control program. It can be used either standalone or together with FLDIGI or third part programs such as the wsjtx. The program supports large number of transceivers from Elecraft, Icom, Kenwood, Ten-Tec, Yaesu, etc. The FLRIG interface changes depending on the degree of CAT support available for the transceiver in use.

Interested users can get detailed information on FLRIG from the following online manual:

http://www.w1hkj.com/files/manuals/US_English/flrig-help.pdf

17.15.1 Installation on Raspberry Pi 5

FLRIG is available on Raspian repository and can easily be downloaded through the Desktop by clicking **Applications Menu → Preferences → Add/Remove Software**. However, the version on the repository may not be the latest version. The latest version (2.0.05) can be installed from the sources as follows (see: https://dl1gkk.com/setup-raspberry-pi-for-ham-radio/ for more information):

- Head to: www.w1hkj.com/files/ and make note of the latest versions of files: **flxmlrpc** and **flrig** by clicking on the files. At the time of writing this book, the latest versions of these files were: 0.1.4 and 2.0.05 respectively. You will be using these versions numbers below.

- Enter the following commands at the Raspberry Pi 5 Console:

  - sudo apt-get install libfltk1.3-dev libjpeg-dev libxft-dev libxinerama-dev libxcursor-dev libsndfile1-dev libsamplerate0-dev portaudio19-dev libpulse-dev libudev-dev
  - cd Downloads
  - wget http://www.w1hkj.com/files/flxmlrpc/flxmlrpc-0.1.4.tar.gz
  - tar –zxvf flxmlrpc-0.1.4.tar.gz
  - cd flxmlrpc-0.1.4
  - ./configure –prefix=/usr/local –enable-static
  - make
  - sudo make install
  - sudo ldconfig
  - cd ~/Downloads
  - wget http://www.w1hkj.com/files/flrig/flrig-2.0.05
  - tar –zxvf flrig-2.0.05.tar.gz
  - cd flrig-2.0.05
  - ./configure –prefix=/usr/local –enable-static
  - make
- sudo make install

- cd ~

17.15.2 Running flrig

Open a terminal session in Desktop and type **flrig** to start the program. Alternatively click on **flrig** in Desktop **Applications Menu → Hamradio → flrig** to start the program. Figure 17.51 shows the flrig startup menu (your startup menu could look slightly different).

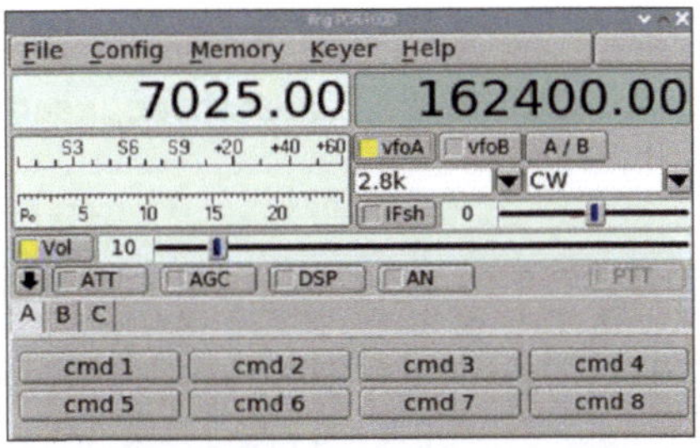

Figure 17.51: flrig startup menu.

Config menu — This menu item has two submenus: **Setup** and **UI** (**U**ser **I**nterface).

Setup — Select **Setup** to configure for your transceiver (option **Xcvr**) as shown in Figure 17.52. Other options in Setup are:

- **Transceiver**: select your transceiver (and serial I/O parameters)
- **TCPIP**: configure for remote TCPIP/serial controlled transceiver
- **PTT**: configure PTT serial ports
- **Polling**: configure transceiver parameters to poll
- **Trace**: configure program execution paths
- **Restore**: read and restore transceiver parameters

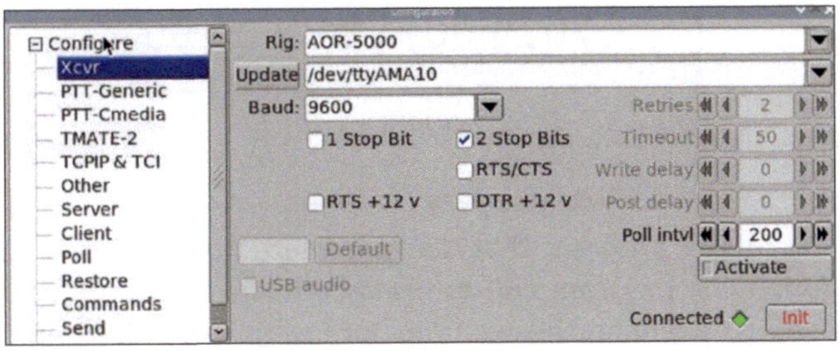

Figure 17.52: Transceiver selection menu.

Note that the Raspberry Pi 5 serial port is named: **/dev/tty/AMA10** whose interface socket is located on the board between the two microHDMI sockets. Select the rig, serial port, baud rate, number of stop bits, poll interval etc. when you select a rig, the default values associated for that rig will be displayed.

There are many options to choose from and a full discussion is beyond the scope of this book. Interested readers should refer to the following online manual and tutorials:

http://www.w1hkj.com/files/manuals/US_English/flrig-help.pdf
http://www.w1hkj.com/FldigiHelp/rig_config_page.html
http://www.w1hkj.com/flrig-help/ft991a_how_to_page.html
http://www.w1hkj.com/flrig-help/ic7300_how_to_page.html

UI (User Interface)

Select this submenu option for meter filters, slider sizing, user interface etc. Figure 17.53 shows the User Interface form.

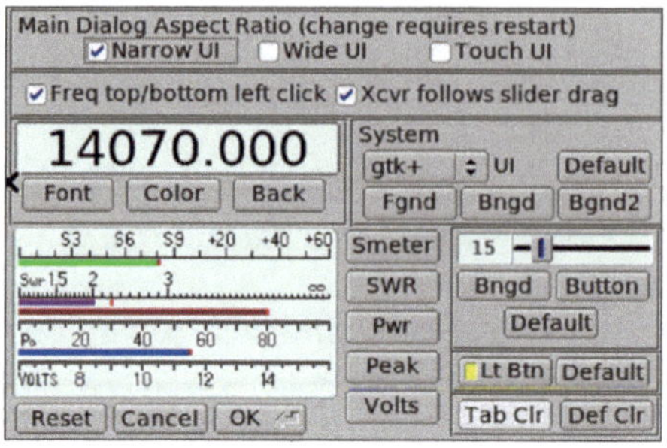

Figure 17.53: User Interface form.

Frequency control

The frequency control display is shown in Figure 17.54. You can use the mouse scroll wheel and left/right buttons to change the values. For example, clicking on the upper half of a digit will increase it, and clicking on the lower end will decrease the number. You can also click on a digit and keep it clicked to change its value. You can click on the A/B button to swap the left and right VFOs.

Figure 17.54: Frequency control display.

17.16 XyGrib

XyGrib is a GRIB file reader program that displays meteorological data in a given location on map in visual form. The program can be used by radio amateurs as well as by sailors, aviators, farmers etc. to examine the weather forecasts. GRIB files are weather forecast files that you can download. A typical file can include a forecast for one day, or for a longer period of time. Of course, shorter forecasts are more accurate.

The basic features of the program are:

- visualization of meteorological data in a given area;
- playing animations of 8-day weather forecasts;
- creating your own weather maps;
- plotting wind speed and direction, pressure, temperature, rain, humidity, snow, total cloud cover, dew point, wave height and high altitude data in visual as well as in text form.

17.16.1 Installation of XyGrib on Raspberry Pi 5

You can install XyGrib from the Desktop by following **Applications Menu → Preferences → Add/Remove Software**. Enter XyGrib in the Search box and press Enter (Figure 17.55). Click **Apply** to install the program and the maps and then click **OK** at the end.

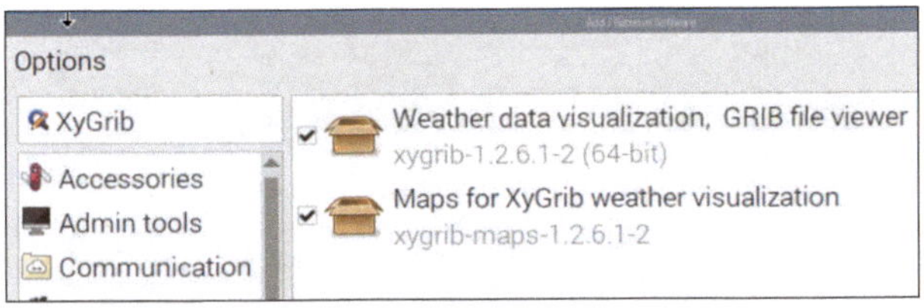

Figure 17.55: Installing XyGrib.

17.16.2 Running XyGrib

To start the program, open a terminal in Desktop mode and enter the command **XyGrib**. Figure 17.56 shows the start up screen where a map of the world is displayed. You can zoom in/out, move left/right by clicking the buttons on the map. By moving the mouse cursor over the map you will see that the geographical coordinates where the mouse pointer is displayed at the top left hand side.

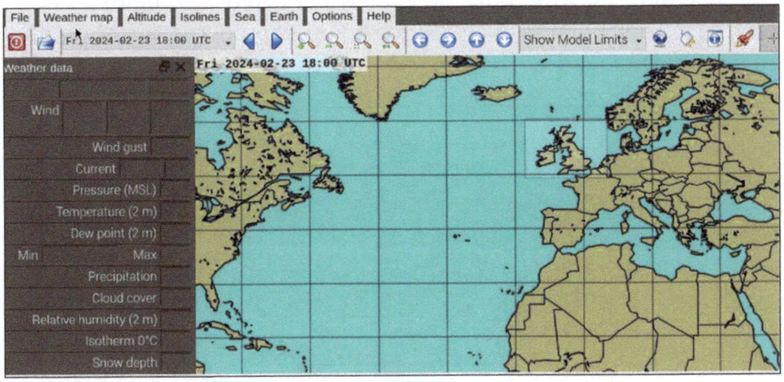

Figure 17.56: XyGrib startup screen.

An example run of the program is given below step-by-step where the UK is selected and the forecasted wind data is plotted on the map.

- Move the mouse pointer over UK and select the UK on the map.
- Zoom-in to the UK.
- Select **File → Download GRIB.**
- Select **Wind** under tab **Surface Data** (Figure 17.57).

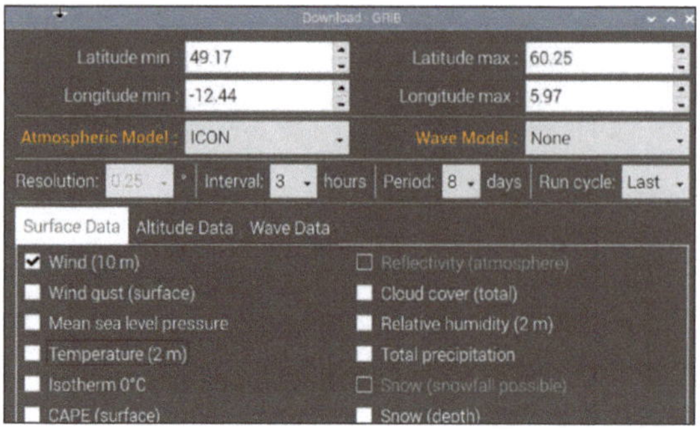

Figure 17.57: Select Wind.

- Click **Download** and save the file to disk.

- You should see the wind data on the map as shown in Figure 17.58. As you move the cursor pointer over the map you should see the wind direction and wind speed displayed at the top left part of the screen.

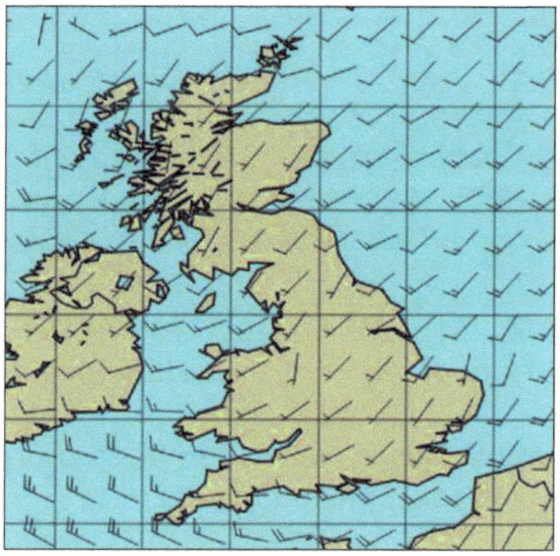

Figure 17.58: Showing the wind data.

- You can change the time or day of the forecast by clicking at the top left part of the screen where it displays the data and time and UTC (Figure 17.59).

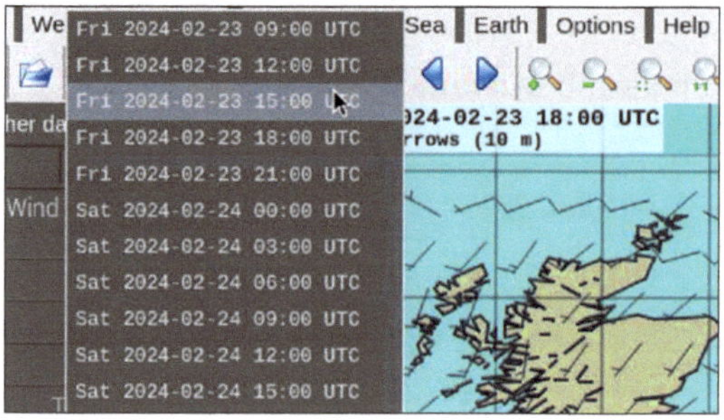

Figure 17.59 Change the time or day of the forecast.

- Click on the **rocketship icon** at the top right hand to animate the display. You should see the time advancing and the state of the wind as the time is changing (Figure 17.60 where only part of the display is shown).

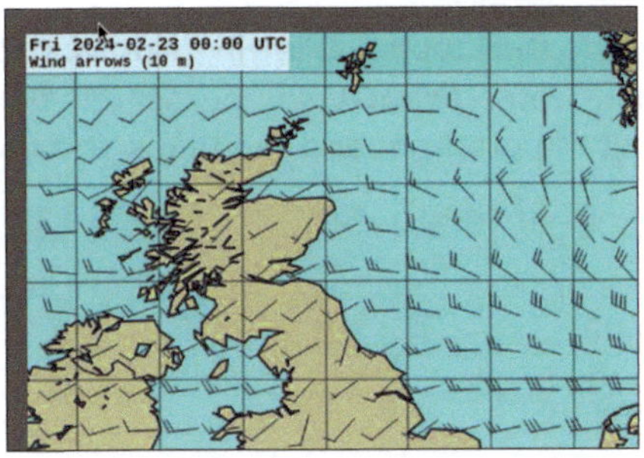

Figure 17.60: Animation of the display.

- Right-click on the top left corner of the display where the date and time are displayed. Select **Meteotable**. In this example, the total cloud cover and precipitation were selected. Figure 17.61 shows the data as a table. The current time is displayed with a yellow background. You can move the table with the slider at the bottom to the right to view the forecast for later days and times.

Location: 60°13'05"N 011°55'56"W
Reference date: DWD-ICON-Global: Fri 2024-02-23 00:00 UTC

	Fri 2024-02-23						
Sun Moon	↑ 08:05 UTC ↓ 17:55 UTC ↑ 16:26 UTC ↓ 08:37 UTC 98%						
	03:00 UTC	06:00 UTC	09:00 UTC	12:00 UTC	15:00 UTC	18:00 UTC	21:00 UTC
Wind (10 m)							
Cloud cover							
Precipitation	0.03 mm/h	0.27 mm/h	0.22 mm/h	0.11 mm/h	0.08 mm/h	0.06 mm/h	0.15 mm/h
Temperature (2 m)							
Gap temp-dew point (2 m)							
Isotherm 0°C							
Pressure (MSL)							

Figure 17.61: Tabulating the total cloud cover and precipitation.

Detailed information on XyGrib can be obtained from the following websites:

https://live.osgeo.org/en/quickstart/xygrib_quickstart.html
https://navigationlaptops.com/xygrib-tutorial-free-detailed-marine-weather/
https://www.youtube.com/playlist?list=PLx1XvLdpAhGA4om2oDuBIsTAUzqdpPSUg

17.17 FreeDV

FreeDV allows any SSB radio to be used for low bit rate digital voice. It is a GUI-based program. FreeDV was built by an international team of radio amateurs working together to design and test the system, and then releasing it under GNU Public License. In FreeDV, speech is compressed down to 1400 bits/s and then modulated onto a 1100 Hz signal which is sent to a microphone input of a SSB radio. At the receiver side, the signal is picked up by an SSB radio and then demodulated and decoded by FreeDV.

Using FreeDV, communication is readable down to 2 dB S/N and very long distance contact is possible with only a few watts of RF power.

17.17.1 Installation on Raspberry Pi 5

You can install FreeDV from the Desktop by following **Applications Menu → Preferences → Add/Remove Software**. Enter **FreeDV** in the search box and press Enter (Figure 17.62). Click **Apply** to install the program and the maps and then click **OK** at the end.

Figure 17.62: Installing FreeDV.

17.7.2 Running FreeDV

You can run FreeDV from the Desktop by clicking **Applications Menu → Hamradio** and then **FreeDV**. When the program is run for the first time, you are required to enter the Radio and Audio details. Figure 17.63 shows the startup screen.

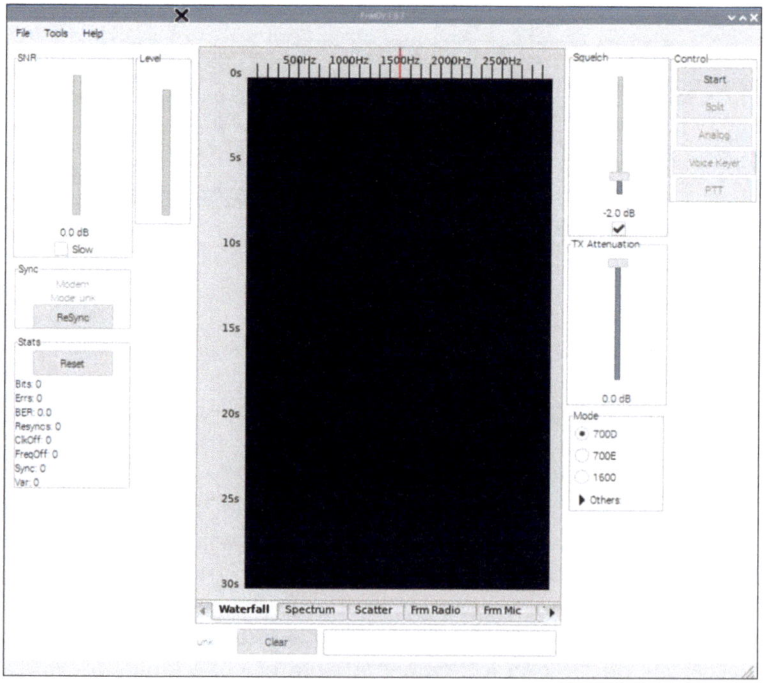

Figure 17.63: FreeDV startup screen.

Detailed information on FreeDV can be obtained from the following websites:

https://www.g0hwc.com/other_digital.html
https://freedv.org/
https://ei7gl.blogspot.com/2019/11/freedv-digital-voice-mode-for-hf-bands.html

17.18 Qtel (EchoLink)

Qtel (EchoLink Client) is an amateur radio system that allows amateurs to communicate with other amateurs using voice over IP (VoIP) technology on the Internet. It is a computer-based system. Using this program, amateurs can communicate reliably worldwide. It is similar to other VoIP applications such as the Skype, but with the ability to communicate to an amateur radio station's transceiver. An *EchoLink node* is an active EchoLink station with a transceiver attached. Any low-powered amateur radio transceiver which has contact with a local node can use the Internet connection of that station to send its transmission via VoIP to any other active node anywhere on Earth. EchoLink is also available on smartphones (e.g. iPhone, Android etc). A computer with a microphone or a transceiver can be used to link to EchoLink.

EchoLink was first developed in 2002 by Jonathan Taylor (K1RFD). Today, there are over 200,000 registered Qtel (EchoLink) users in over 151 countries.

17.18.1 Qtel installation on Raspberry Pi

You can install Qtel from the Desktop by following **Applications Menu → Preferences → Add/Remove Software**. Enter **qtel** in the search box and select **Graphical Client for EchoLink Protocol**, press Enter. Click **Apply** to install the program and the maps and then click **OK** at the end.

17.18.2 Running the program

You can run the FreeDV from the Desktop by clicking **Applications Menu → Hamradio** and then **Qtel**. When the program is run for the first time, you are required to enter the Radio and Audio details. Figure 17.64 shows the startup screen (part of the screen is shown).

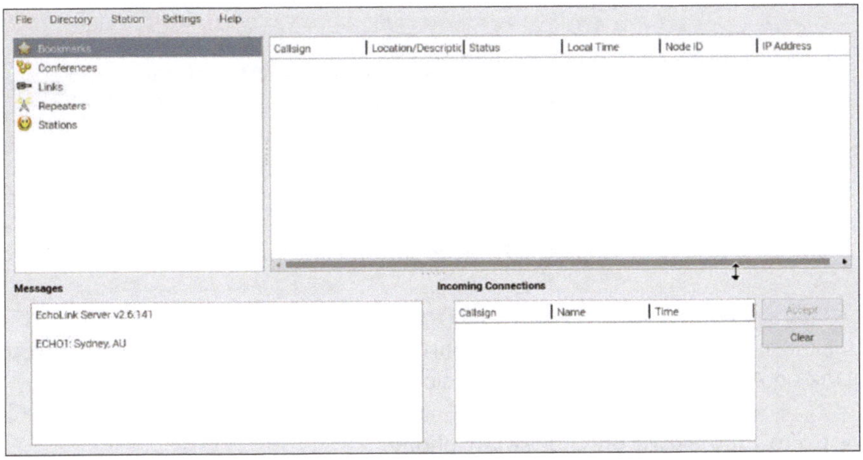

Figure 17.64: Qtel (EchoLink Client) startup screen.

You have to register at http:/www.echolink.org/validation and validate that you are a licenced radio amateur before you can use the EchoLink program. This involves entering your callsign, address, email address, and sending a scanned copy of your radio amateur certificate to the given web link. The validation process takes no more than 24 hours and you will be notified of successful validation by an email. After this, enter your password in **Qtel Settings** in the **User Info** section, and start using the program to communicate with other radio amateurs. If you forget your password, you can send for a new password by searching Google for "echolink forgot password". Enter your callsign and your email address. A password reset link will be sent to your email where you can create a new password.

Figure 17.65 shows the program screen after the radio amateur license validation. At the left pane you have the menu options: **Conferences**, **Links**, **Repeaters**, and **Stations**. You can, for example, select the Conferences and select a conference room for communication. Double-click on the selected conference room and then click on **Connect**. Other useful information, such as the station location/description, status, local time, Node ID, and IP address are also displayed.

Note that EchoLink uses UDP ports 5198 and 5199 for incoming requests, and TCP port 5200 for outgoing requests. You should make sure that these ports are enabled on your Wi-Fi router to direct all incoming data on these two ports and assign them to the Raspberry Pi 5. This can be done using **Forwarding** on your router, or by doing **Port Triggering** if you wish to use EchoLink on several different computers.

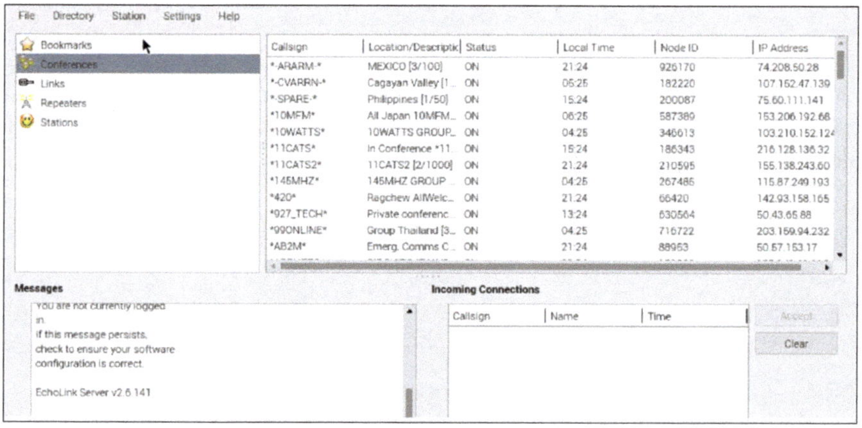

Figure 17.65: Program screen.

17.18.3 Using EchoLink on smart phones

You can also use EchoLink apps on your smartphone (iPhone or Android). The steps to set up EchoLink on Android phones are given below:

- Go to **Play Store** on your smart phone.
- Search for echolink apps (Figure 17.66) and install it.

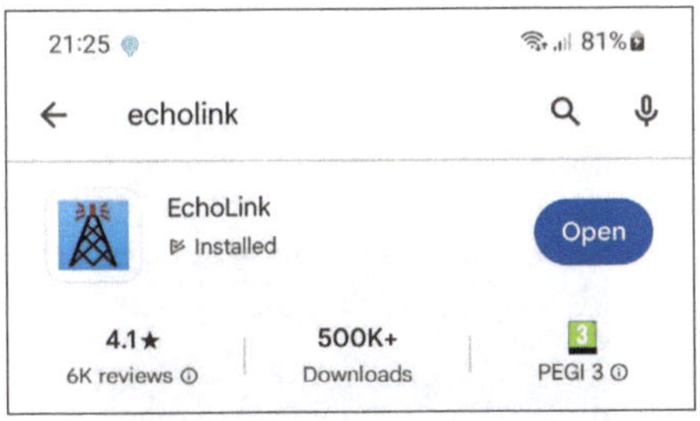

Figure 17.66: EchoLink in Olay Store.

- Start the apps and enter your password.
- Figure 17.67 shows the start up screen.

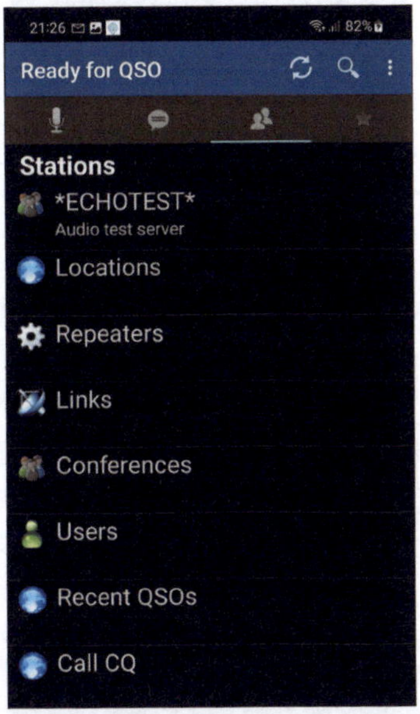

Figure 17.67: EchoLink start up screen on Android smartphone.

Figure 17.68 shows the Conferences submenu. The settings submenu is shown in Figure 17.69.

Figure 17.68: Conferences submenu.

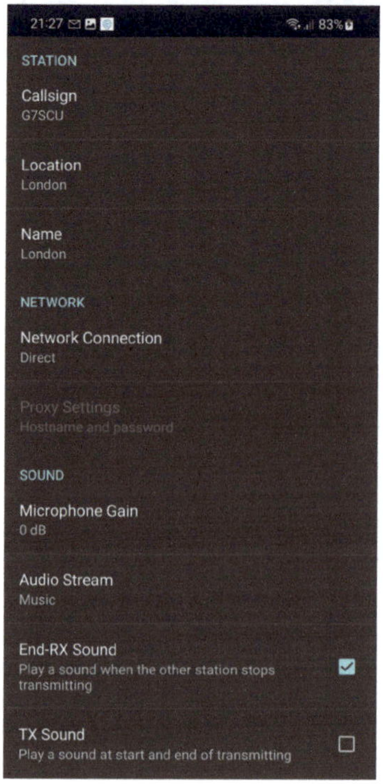

Figure 17.69: Settings submenu.

17.19 XDX (DX-Cluster)

DX Clusters provide real-time information on amateur radio contacts around the world. The clusters collect messages from active radio amateurs and then distribute them to connected participants. The messages contain information such as the callsign, time, frequency, comments, etc. Radio amateurs can use DX Clusters to acquire information about activities on the amateur radio bands. Whenever a DX Cluster receives a message, it is sent to all other DX Clusters and to connected participants.

17.19.1 Installation on Raspberry Pi 5

You can install Qtel from the Desktop by following **Applications Menu → Preferences → Add/Remove Software**. Enter **xdx** in the Search box and select **DX-cluster tcp/ip client for amateur radio**, then press Enter. Click **Apply** to install the program and the maps and then click **OK** in closing.

17.19.2 Running the program

You can run xdx from the Desktop by clicking **Applications Menu → Hamradio** and then **xdx**. Click **Settings → Preferences** and enter your callsign, web browser, etc., under the **General** tab as shown in Figure 17.70. Select **Output** under the **Settings → Preferences** menu to direct the output and also select the fields to be displayed. Eight function boxes are provided that can be set to any required valid **xdx** commands (you can remove the function keys bar from the **Settings** menu if you wish).

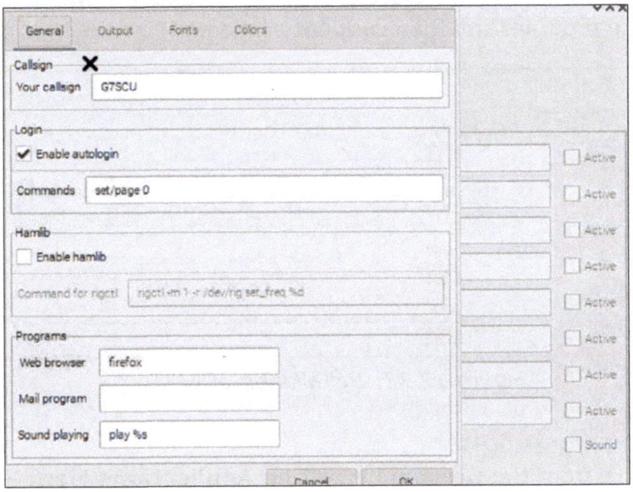

Figure 17.70: Preferences menu.

Click **Host** and enter a DX Cluster host name and port number. In this example, the host name **dxspider.co.uk** was used with the port number set to **7300**. There are many DX Cluster nodes available, and they all connect to the same network and provide similar services. Figure 17.71 shows a typical display from the program.

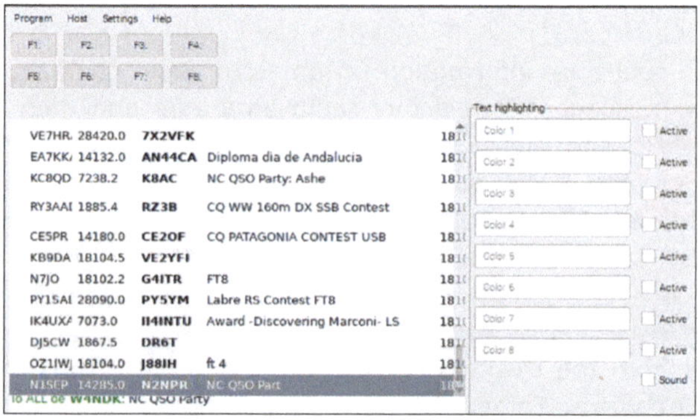

Figure 17.71: ßTypical display from xdx.

17.20 WSJT-X

WSJT-X is a program developed for amateur radio communication using very weak signals. The program was initially developed by Bill Somerville (G4WJS), Steve Franke (K9AN), and Nico Palermo (IV3NWV) between 2001 and 2016. The program offers 11 different protocols, including: FST4, FT4, FT8, JT4, JT9, JT65, Q65, MSK144, WSPR, FST4W, and Echo.

17.20.1 Installation on Raspberry Pi 5

You can install WSJT-X from the Desktop by following **Applications Menu → Preferences → Add/Remove Software**. Enter **wsjt** in the search box as shown in Figure 17.72. Click **Apply** to install the program and then click **OK** at the end.

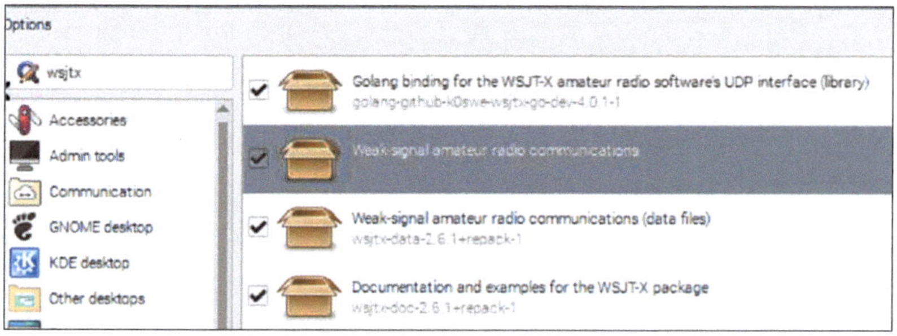

Figure 17.72: Installation of WSJT-X.

17.20.2 Running the program

You can run WSJT-X from the Desktop by clicking **Applications Menu → Hamradio** and then **wstjx**. Figure 17.73 shows the start up screen of the program (only the upper part of the screen is shown).

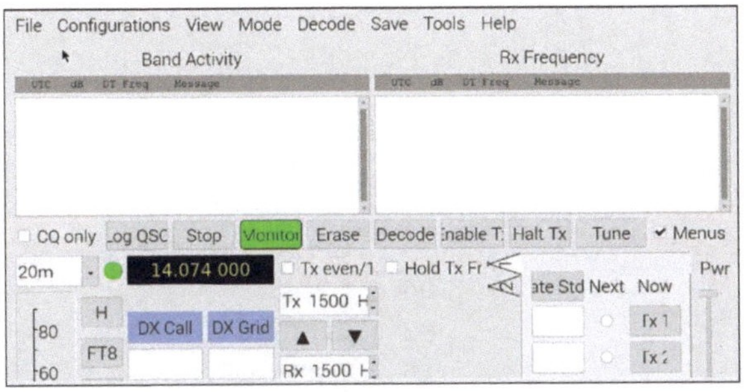

Figure 17.73: WSJT-X launch screen.

The use of the WSJT-R program is unfortunately beyond the scope of this book. Interested reader will find the following websites useful:

https://wsjt.sourceforge.io/wsjtx-doc/wsjtx-main-2.6.0.pdf
https://wsjt.sourceforge.io/wsjtx-doc/wsjtx-main-2.6.1.html
https://en.wikipedia.org/wiki/WSJT_(amateur_radio_software)
https://wsjt.sourceforge.io/wsjtx.html

Index

U

V

W

X